What Physics says

Science is made of facts, just as a house is made of stones, but an accumulation of facts is no more a science than a pile of stones is a house
(Henri Poincaré)

Physics seeks what is simple and invisible behind what is complicated and visible
(Jean Perrin)

Featuring: Bear, Penguin, Seagull, Myrtle the sheep, Barky the Dog ... and Professor Eifel!

Contents

The authors

This book was produced by two scientists, François Louchet and Serge Castel.

François Louchet obtained an engineering degree at Ecole des Mines, received the "agrégation" degree in Physics, and a PhD in Condensed Matter Physics. He was a University Professor at the "Institut National Polytechnique de Grenoble", and carried out his research work at the Thermodynamics and Physico-chemical Metallurgy Laboratory, and at the Grenoble Glaciology and Environmental Geophysics Laboratory. He ended his career as an "exceptional class" Professor Emeritus, and continues his research activity in Theoretical Physics and Geophysics.

He has also been a Visiting Professor at the Ecole Polytechnique Fédérale de Lausanne (EPFL), Guest Scientist at Exeter University (UK), at the Los Alamos National Laboratory (Department of Energy, USA), at McMaster University (Hamilton, Canada), at Charles University, Prague (Czech Republic), and at the Balseiro Institute (Comision Nacional de Energia Atómica, Bariloche, Argentina). He also was OCMR Distinguished Lecturer at Hamilton (McMaster), Toronto and Kingston Universities (Canada).

He focused his research activity on the theory of evolution of complex systems, applying such mathematical tools to dislocation avalanches in crystalline solids, snow avalanches, propagation of contagious epidemics, and climate.

Serge Castel is a chemical engineer. After studying in Strasbourg (France) and Delft (Netherlands), he pursued a career in major international chemical companies, holding positions in process development, production, operational excellence and change management. He also is a "Lean Six Sigma Master Black Belt".

In his spare time, he draws and paints in watercolors. He practices satirical drawing and keeps a chronicle of municipal life in a small town in Alsace.

Dedicace

We dedicate this book to the memory of Hubert Reeves and Joe Blanc.

When one of us (FL) published his first paper on climate change in 2016, he sent a copy to the astrophysicist Hubert Reeves. Hubert responded with a post in his journal Humanité & Biodiversité (http://www.humanite-biodiversite.fr/article/hubert-reeves-s-interroge), in which he emphasized the importance of this approach in terms of critical transition and tipping point, and warned of the imminence of this tipping point and the amplitude of the associated temperature jump. Thank you, Hubert. How sad that we will no longer be able to share and debate with you.

Joe Blanc was a researcher in theoretical Physical Chemistry in Princeton, and a close friend of FL with whom he had numerous scientific discussions, more particularly in fundamental Thermodynamics, stability of systems, and more recently on Climate. Thank you for all these fruitful exchanges and for your long lasting friendship. We miss you very much, Joe.

Foreword

What's going on?

I don't know what's going on
Says the Earth: I feel dizzy.
Have I been spinning too much in space?
Or drinking too many bitter liquors?
Red mud, acid rain,
Verdigris in Rhine gold,
Herbicide, pesticides,
These are malignant poisons!
It's so strong that I'm losing my mind,
My Poles are crooked,
My head is drunk:
I'm seeing the Universe upside down!
Remembering my apple roundness
In the beginning of time,
Just before man's teeth
Got eagerly stuck in it.
...

The first four stanzas of Marc Alyn's poem "I don't know what's going on"
Companions of the Marjoram. Ed. de l'Atelier. 1986

1
Introduction

"He who knows does not speak, he who speaks does not know"

Forewarned by this dreadful Lao-Tzu maxim, we cowardly decided...

... to write and draw.

A new and original graphic book on climate change? Yes, indeed. **New,** because our approach is based on theoretical physics, whose tools have only recently begun to be applied to climate change. **Original** because we rely on the Theory of Dynamical Systems to take us off the beaten path towards new ideas about climate change. **Graphic**, because this format is probably the best way to show that the concepts we are explaining are simpler than they may seem at first glance.

While this book is intended for a broad audience, it is grounded in rigorous scientific arguments. We use simple thought experiments and diagrams throughout, which we believe help to demystify the science of climate change without betraying our commitment to scientific rigor. The conversations of our story's characters add a touch of humor and levity to a subject which is no laughing matter.

Along with Bear, Penguin, Myrtle the Sheep, Barky the Dog, the Shepherd and Professor Eifel, we will discuss the current state of our climate, how climate change works, and to what extent we can predict its evolution. Rather than relying on sophisticated calculations to predict what lies ahead, the Theory of Dynamical Systems instructs us to listen to what the planet itself is telling us. Using the language of more and more frequent and powerful "extreme" climatic events, the Earth is indeed warning us that we are irreversibly heading towards a much warmer state**, a leap into the unknown.**

Such warning signals, called 'pre-critical fluctuations' in Theoretical Physics, are unfortunately all too often ignored in other fields of Science. Beyond warning us of an incoming tipping point, they can be used to estimate its date of occurrence. On the other hand, it can be shown that, in the vicinity of the tipping point, which is a mathematical singularity, numerical simulations, no matter how sophisticated, cease to be reliable. This explains their excessive optimism and shows that a major tipping event is likely to occur in the next few years, Such a result invalidates in advance any political or technical decision that would have no practical effect before that date. We are in a situation of extreme emergency.

At the same time, 'as the story unfolds', we will attempt to critically unravel a number of preconceived ideas, some of which being tenacious, and all the more widespread as they are inaccurate.

For those of you curious enough to go further, which we hope you will, we have included a list of scientific references at the end of the book (which is unusual for a comic!), most of which are freely available on the Internet.

The Unlikely Encounter of
the Polar Bear, the Antarctic Penguin and Professor Eifel

North Pole
South Pole
Hi Bear !
Hi Penguin !
Oh, you saw over there?
Looks like Professor Eifel.

2
Where the hell is climate change taking us ?

We must always tell what we see, but above all, and this is more difficult, we must always see what we see.
(Charles Péguy)

Winter just isn't what it used to be. We started taking refuge in the North Pole, because experts told us it was the only place in the world where there was no north wind. But it was still pretty damn cold..

You're very lucky up there. In Antarctica we have nothing but the north wind. Maybe that's why the continent is so cold ! But actually, I'm wondering whether all this north wind story was nothing but a stupid joke. One day, a nice guy named Mathieu came up here to see what was going on far from the coastline, on the ice cap, very close to the South Pole, where everyone thought it was freezing cold . Well, it wasn't (Casado et al. 2023)! He published a paper saying that it was warming up awfully fast, almost as fast as (or even more than) on the coast.

Same here! We don't even bother with the North Pole anymore, because everything has warmed up so much, even in winter! Instead, we still enjoy hunting at the ice pack, where we can catch seals. They are easier to get than the walruses, who live in groups and have fearsome tusks.

True, it's not a joke ! And we're starving around here. Our babies hardly have time to learn to swim before orcas start in on them.

Yes, quite worrying. At the moment, scientists are keeping a close eye on the huge Thwaites Glacier, as large as the United Kingdom, which flows into the Amundsen Sea , west of the Antarctic Peninsula. It heats up so much that it "softens", like butter, and flows faster and faster towards the ocean

(Alley et al. 2021, Pettit et al., 2021, Wild C.T. et al., 2022).

Oh well, I got it. Is that why the bottom end of the glacier is now floating on the ocean, why it is falling apart, why there are more and more icebergs in the sea?

Yes, exactly.

Some people believe that there are more icebergs because it's getting colder. But it's just the opposite!

Funny !

No, not really, and that's a big deal. As long as it is on water, this floating ice cools it by melting like an ice cube in a glass and is renewed by glacier flow. This very cold water is then carried along by currents across world's oceans, slowing down their warming. It's the same with Greenland's glaciers.

But one day, by dint of melting, the glaciers in Antarctica, Patagonian Andes and Greenland will have retreated to dry land, and will no longer be dumping ice into the oceans, but simply melted water. And the whole ocean mass will then warm up much, much faster, just like a glass of water whose last ice cube has melted!

And the Arctic? It's getting hot in here!

Yes of course. Except that in the Arctic, it is mostly sea ice, made of frozen seawater, which is already floating freely on the ocean. It is as white as snow and reflects back to the sky most of the solar radiation it receives, like snow and ice of the Antarctic continent. It's called albedo. But when it melts, it is replaced by much darker seawater, which absorbs solar radiation and heats up faster.

And isn't that likely to raise sea levels ?

No, not at all in this case. What you're saying, Bear, is true for melting of glaciers that lie on land, which could raise sea levels by a few meters or even tens of meters across the globe, drown coasts and even entire regions, but not for ice that floats on water.

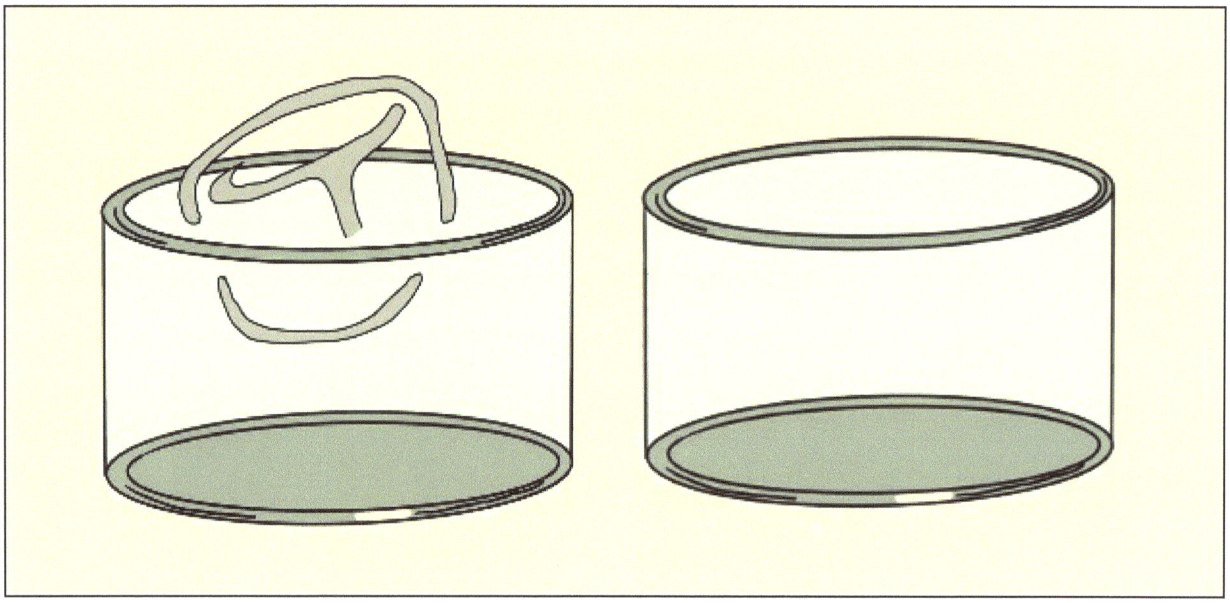

If you want proof, take a glass if you find one in a den of researchers in Svalbard. Put an ice cube in it, fill it with water up to the brim, and let the ice cube melt. You will see that not a drop of water will overflow. This was discovered by Archimedes while taking a bath in Syracuse a long time ago. But water wasn't that cold, and the ice cube sticking out of the water was Archimedes himself!

Yes, but hasn't all of this, global warming, melting ice, rising oceans already happened to Earth long ago, and more than once? Things worked themselves out before. Why are we worrying now?

Yes, there have been ice ages, interglacial eras, and so on. But they happened long ago, when there were no penguins, bears or humans on Earth.

Interesting! But if happened before without humans, why do we think humans are responsible this time?

OK, indeed, That's a point of view we hear sometimes, and it requires a little explanation. Now let's have a brief look on history...

Earth temperature as compared to average of 1960 – 1990 years

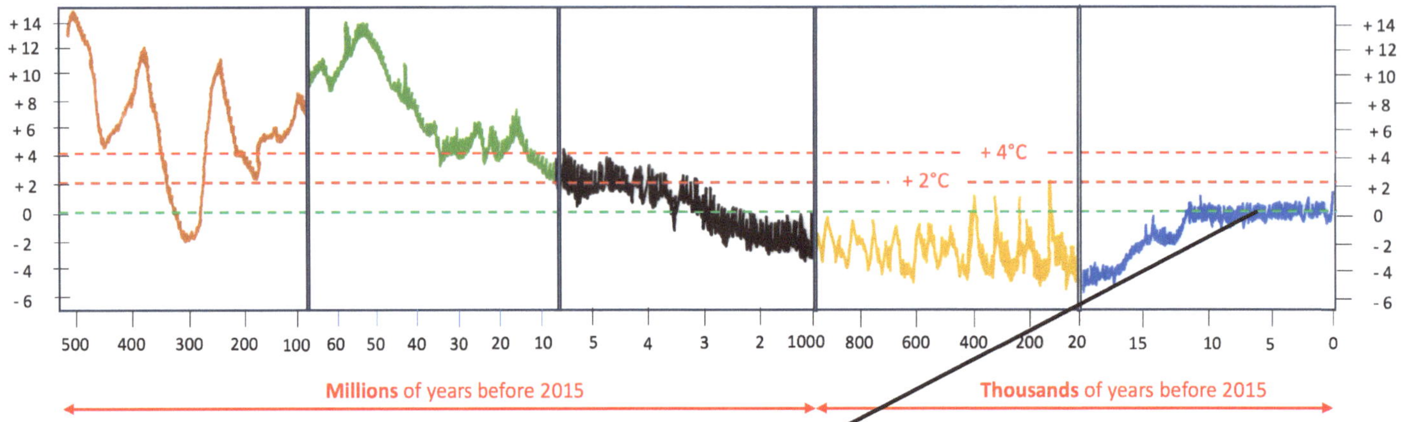

Ref. https://en.wikipedia.org/wiki/Geologic_temperature_record

Earth's temperature has always fluctuated. Here are temperature variations over hundreds of millions of years compared to a reference which is the average temperature in the 1960s - 1990s.

WATCH OUT ! The time scale is not linear! The first three boxes are graduated in million years, and the next two ones in thousand years. And in addition, within these areas, moving from one box on the left to the next one on the right results in zooming in again.

You can see in blue at the tip of my wand that temperature over the last 10,000 or 12,000 years has been remarkably constant, fluctuating less than 0.5 °C .

Due to this stabilization, my human ancestors could gradually evolve from hunter-gatherers to sedentary farmers.

But if we zoomed in even more on the last few years, we would see (as we can guess from the figure) a runaway warming, which has already taken us to nearly 2 °C above what we have experienced in the last 10,000 years, and which is apparently not about to stop, as we shall see later on.

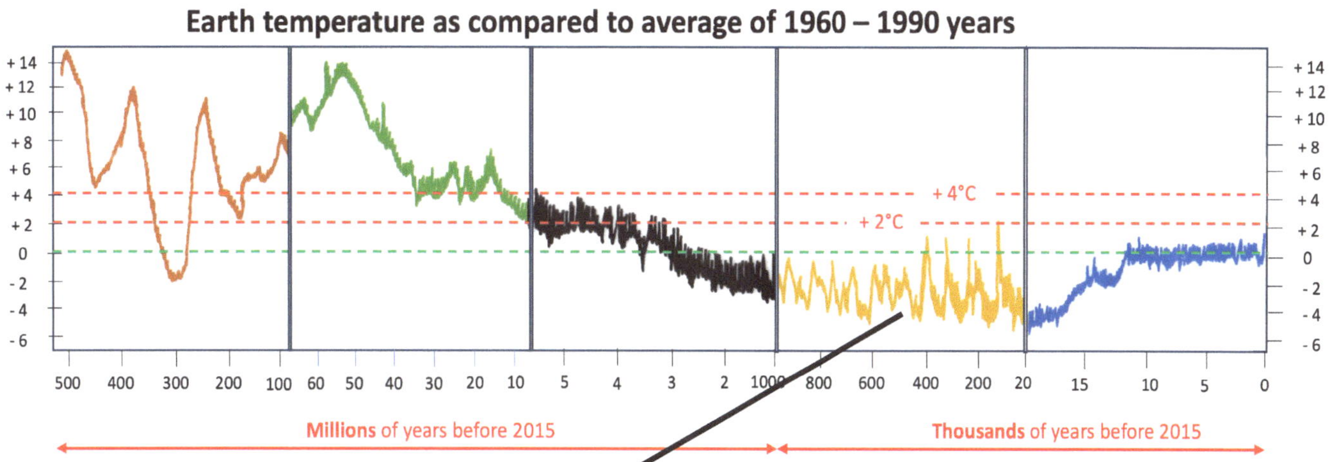

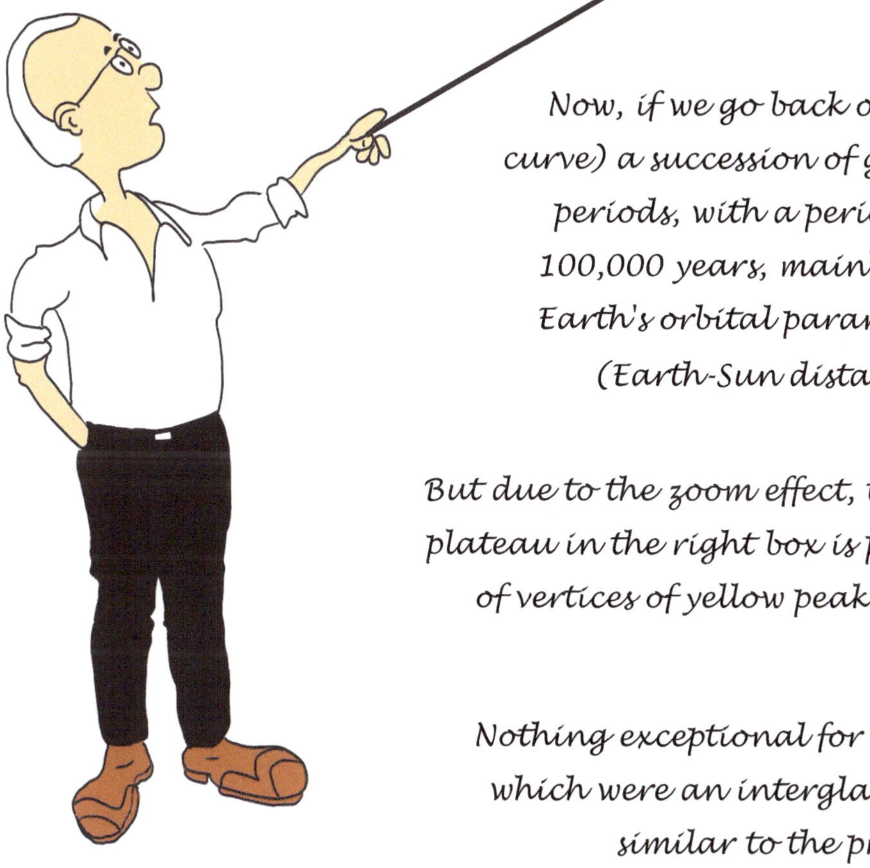

Now, if we go back one box, we see (yellow curve) a succession of glacial and interglacial periods, with a periodicity of the order of 100,000 years, mainly due to variations in Earth's orbital parameters around the Sun (Earth-Sun distance in particular).

But due to the zoom effect, the duration of the blue plateau in the right box is probably similar to that of vertices of yellow peaks in the previous box.

Nothing exceptional for the last 10,000 years, which were an interglacial (warm) period similar to the previous ones.

On the other hand, the current surge in temperatures, which is heading towards 3 or 4°C in the next few years, is an unusual warming as compared to former naturally warm periods! It has nothing to do with variations in orbital parameters, contrary to what is sometimes claimed, and in particular by the warming suddenness (150 years instead of about 50,000 years for warming periods in the "yellow" zone).

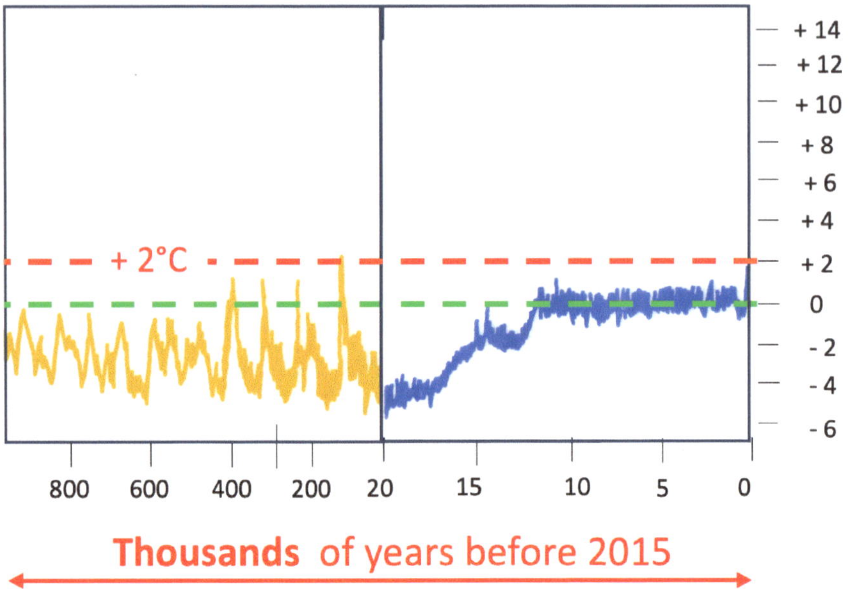

We can see that we have to go more than 130,000 years back to find a temperature 2°C higher than that of our pre-industrial reference.

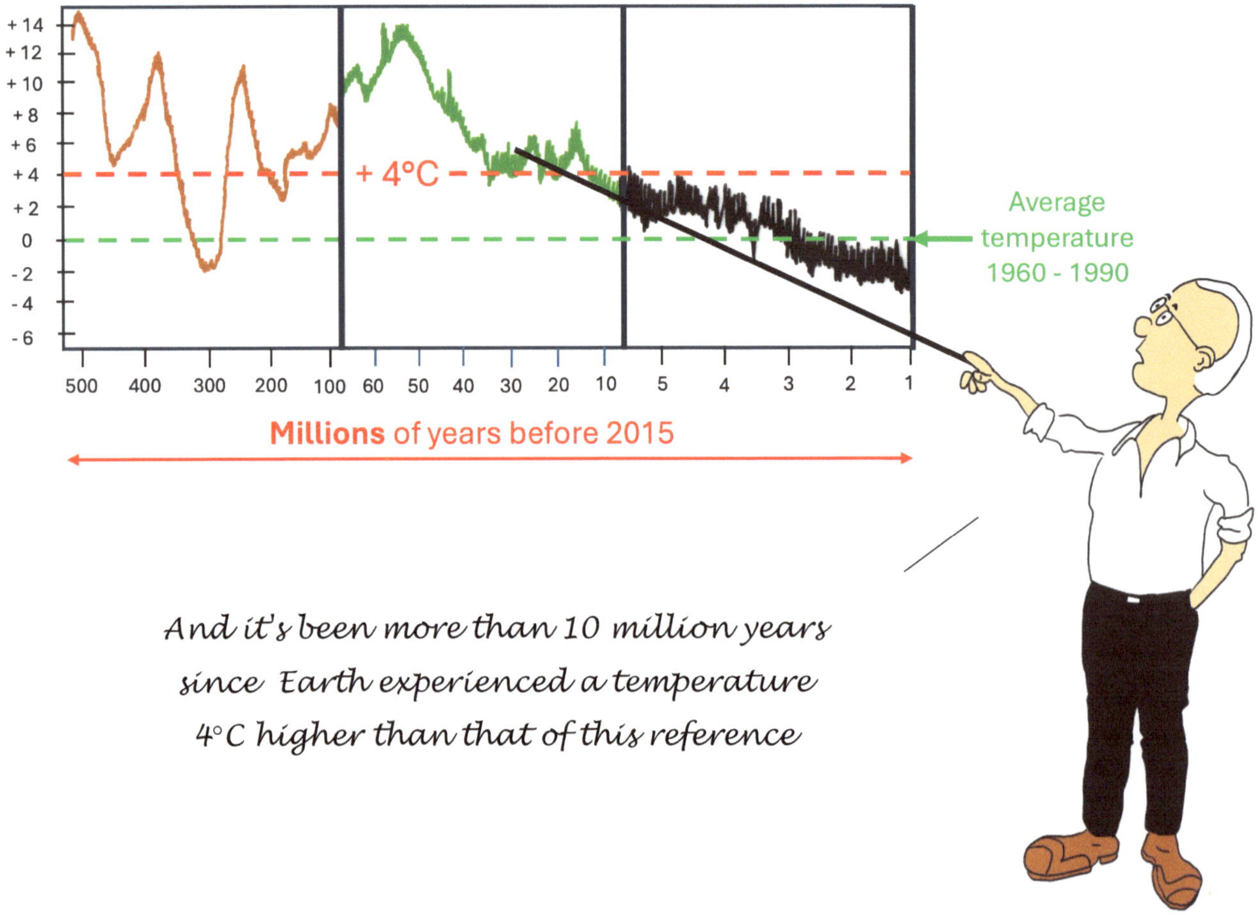

And it's been more than 10 million years since Earth experienced a temperature 4°C higher than that of this reference

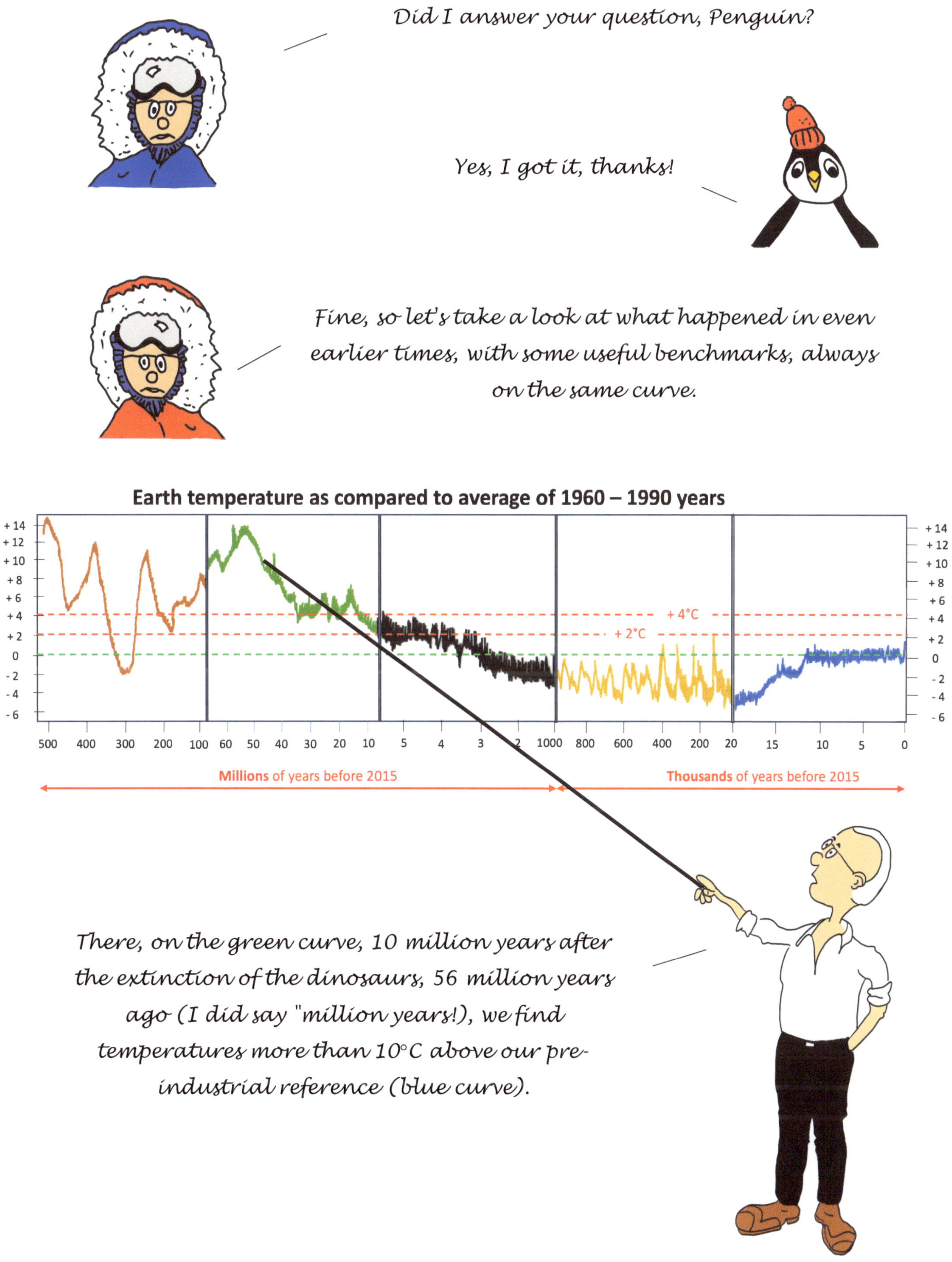

Did I answer your question, Penguin?
Yes, I got it, thanks!
Fine, so let's take a look at what happened in even earlier times, with some useful benchmarks, always on the same curve.
Earth temperature as compared to average of 1960 – 1990 years
+ 14 + 12 + 10 + 8 + 6 + 4 + 2 0 - 2 - 4 - 6
+ 4°C
+ 2°C
500 400 300 200 100 60 50 40 30 20 10 5 4 3 2 1000 800 600 400 200 20 15 10 5 0
Millions of years before 2015
Thousands of years before 2015
There, on the green curve, 10 million years after the extinction of the dinosaurs, 56 million years ago (I did say "million years!), we find temperatures more than 10°C above our pre-industrial reference (blue curve).

This is known as the Paleocene-Eocene Thermal Maximum (PETM) (McInenney et al., 2011). It resulted from a massive increase in the amount of CO_2 produced by intense volcanic activity. It was not significantly different from today's in terms of the amount of CO_2 emitted. But this CO_2 emission was spread over several tens of thousand years, compared to only 150 years today !
But we'll come back to that a little later !

Now, why such variations ?
To understand clearly, I think it's time to take a step back. Let's look at our Earth from space.

3
How does climate work ?

What gets us into trouble is not what we don't know.
It's what we know for sure that just ain't so
(Mark Twain)

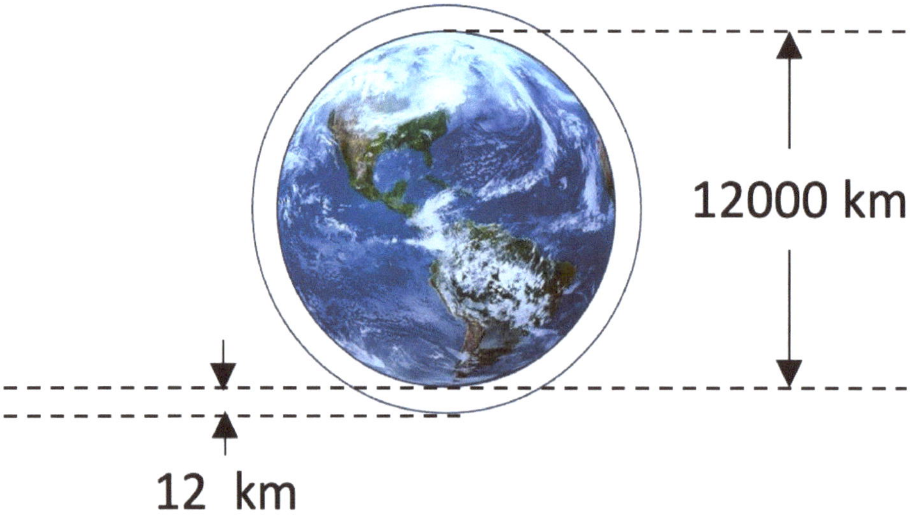

Imagine this. If Earth was 1 m in diameter, the atmosphere would be only 1 mm thick. Incredible !

And yet, this thin and delicate layer protects us from cosmic radiations and allows us to breathe !

*And it also plays another essential role, due to the so-called **GREENHOUSE EFFECT**.*

It is a natural phenomenon and a mandatory condition for the type of life that developed on Earth. Our planet gets all its energy from the sun. Only part of this energy is absorbed by the soil and atmosphere. The rest is sent back into space. Again, this is an albedo effect, largely due to clouds, seen from above!

But in addition, greenhouse gases prevent a significant part of infrared rays emitted by Earth itself due to its temperature (this is called Stefan's law) from being sent into space.

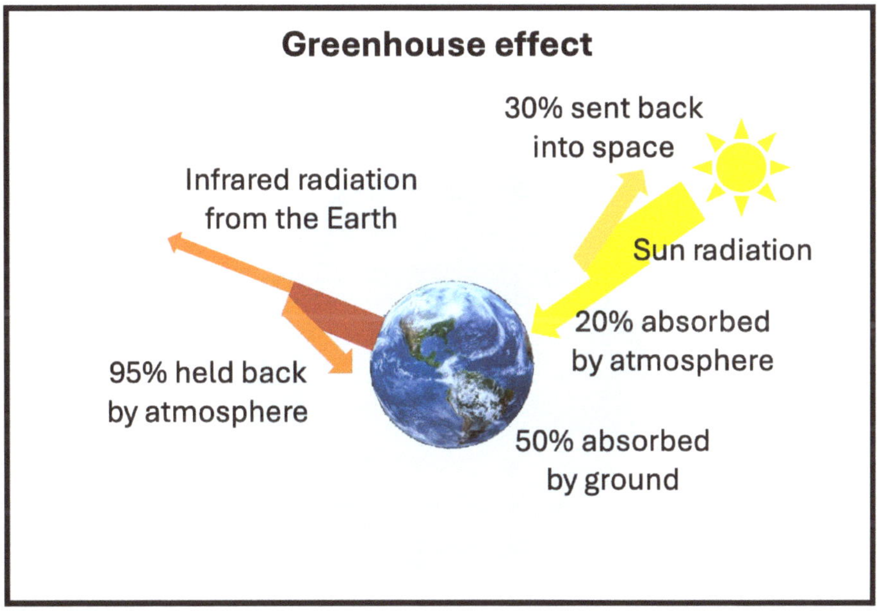

Greenhouse effect on Earth is due to several gases, essentially:

1. **Carbon dioxide CO_2**, which we hear a lot about, is the main culprit of current warming.

2. **Methane CH_4**, about 25 times more efficient than CO_2, but much less abundant and disappearing quickly.

3. **Nitrogen oxides**, the famous **NO_x**, but which also degrade relatively fast.

4. But also, and this is less well known, **water vapor** !

Water vapor?
Did you say water vapor ???

Yes, I confirm. Water vapor is a powerful greenhouse gas. Without it, Earth would be an icy desert, just like Mars, which experiences temperatures between -150°C and +35°C, depending on the season, latitude and hour in the day. But, on the contrary, the reason Earth is not a furnace is because water is present in its 3 states, solid, liquid and vapor, which are in equilibrium with each other.

In equilibrium ?
What do you mean ?

What I mean is that the atmosphere is often saturated with water vapor...

… and any possible extra water vapor injected into the air by human activity (or anything else) is automatically compensated by precipitation (rain or snow), bringing the system back into equilibrium.
It regulates itself!
This is a key mechanism in the selection of a variety of life types, that could develop, mutate and adapt to each other and to climate fluctuations over the hundreds of million years that preceded us, up to what we know today.

And it's not the same thing for CO_2???

No, not at all, alas!
Because CO_2 naturally exists on Earth only as gas. There is no precipitation in the form of rain or dry ice, because the climate is already too warm for that. Its concentration was pretty much regulated by photosynthesis before the industrial era. But current human activity is producing it at such a rate that Nature no longer has time to react.
And that's why temperature is soaring.

But... if it gets too hot, couldn't we fly (as I do), but to Mars, where it's much colder as you said ?

Yes, some people have thought about it! But that's pure madness! Even if Earth became too hot for us , fragile oxygen-dependent earthlings, the climate in Mars would be far worse.

There is no hope of creating "livable" temperatures there for us, because of the virtual absence of water vapor, and the impossibility of introducing even a little amount of our precious oxygen, which vanishing of magnetic field (4 billion years ago), weak gravity and solar wind would soon send into space with no hope of return.

But then, what to do ?

Most of the CO_2 we add comes from fossil fuel burning and from deforestation. It is the main culprit in the increase in greenhouse effect. And we keep adding more and more !!

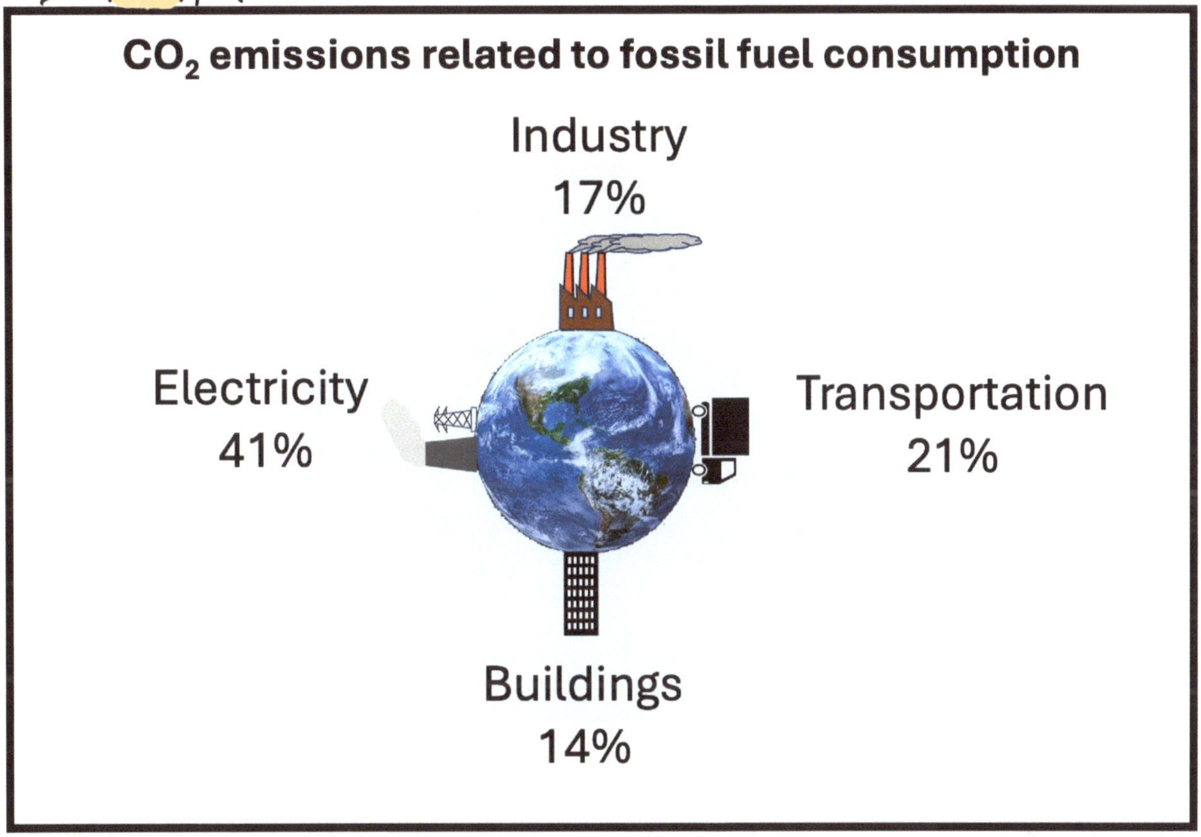

Since the beginning of the industrial era, it has almost doubled, and the pace is accelerating again and again !

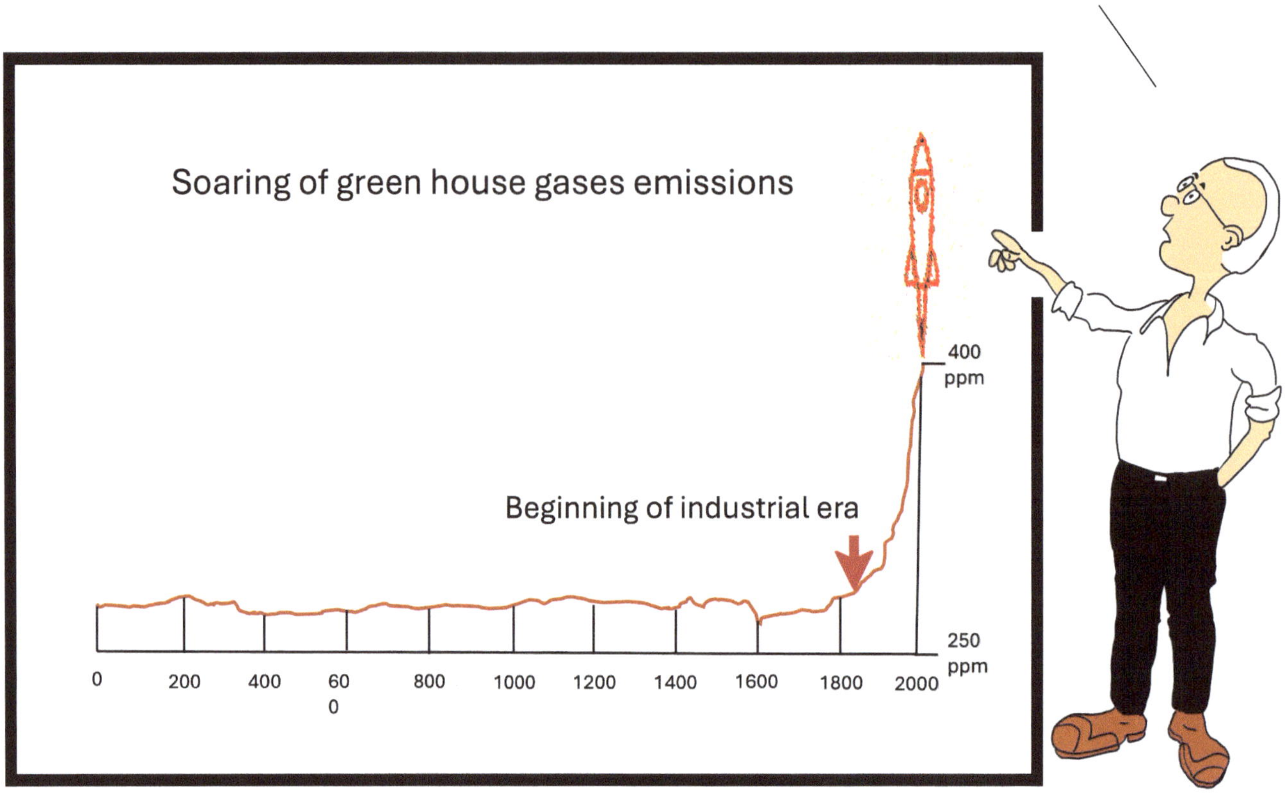

Atmosphere green house gases content in ppm (parts per million)
From: https://www.youtube.com/watch?v=7vopu9X2mcc
https://nas-sites.org/americasclimatechoices/more-resources-on-climate-change

4
What to expect now ?

It is because science is not sure of anything
that it is constantly making progress
(R. Massain, Physics and Physicists)

4.1 Stretching or tipping ?

False Hopes Are More Dangerous Than Fears
(John R.R. Tolkien)

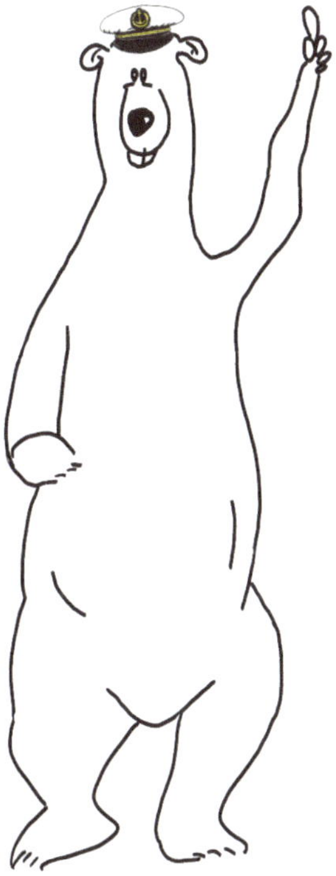

Pretty scary !
But we should be able to reverse it,
if we want to, right ?

Not obvious at all! You
know Labrador quite
well, Bear, I presume?
Let's go for a canoe ride
on the Churchill River.

What a nice, quiet river!

Well, it starts shaking quite a bit now. If it goes on,
the canoe will start taking in water.
But hey, don't worry, I'll bail it out!

Help! Rescue! Nobody warned me!!!

Scary!
Never seen anything like this before!
"Who would have guessed ???

Well, I love jumping into water. Not afraid of the cold ! But that, Brrr!! I can't fly!! How do you know if you're approaching a waterfall ?

That's the right question. If you pull on a rubber band, it stretches... If you pull harder and harder, it stretches more and more. But if you keep stretching, all of a sudden, it snaps!

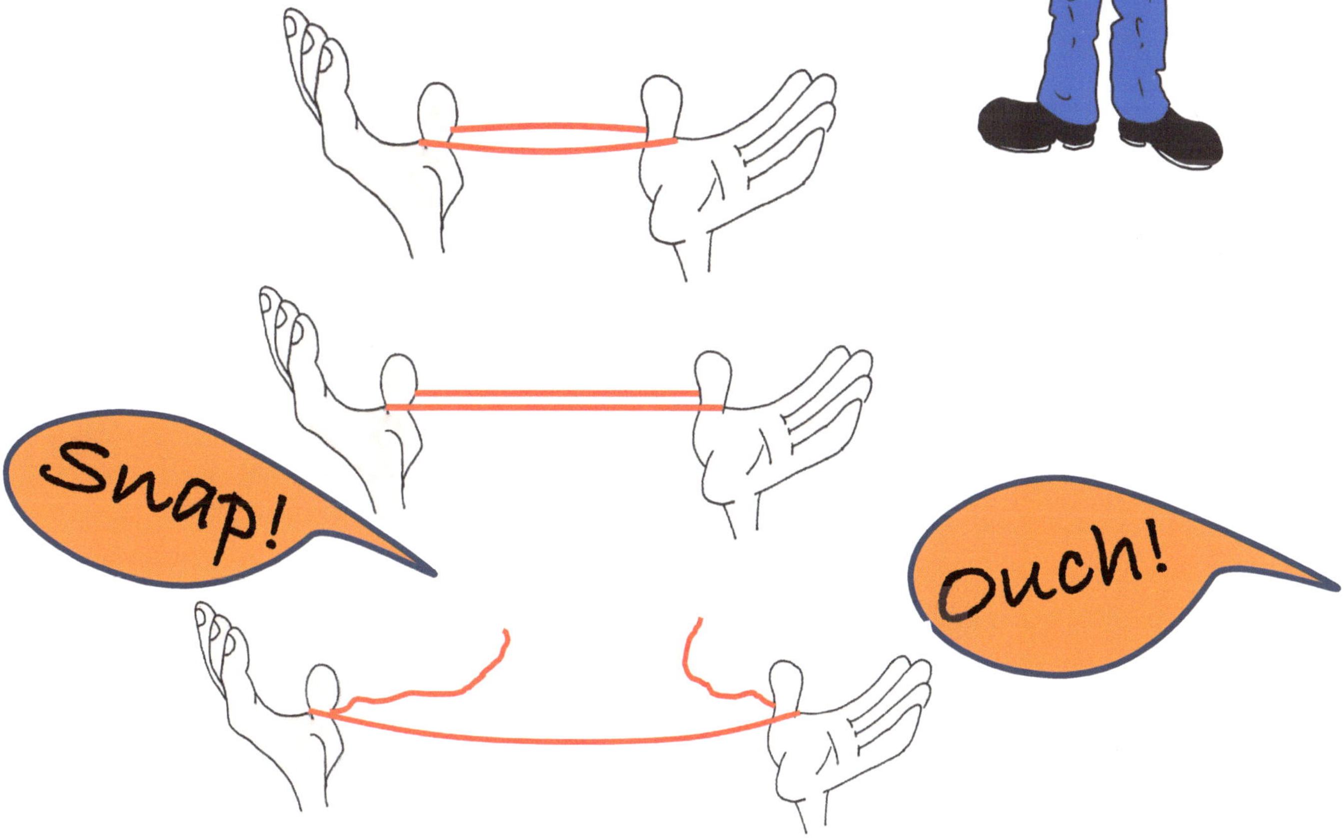

And there are warning signs. as we'll see now.

Let's try tearing up the piece of burlap below.
I start stretching vertically for instance. A first fiber fails, that makes the tear somewhat bigger.
I go on stretching harder and harder, which breaks 3 other fibers, then 10, then 30, like so many cascades of dominoes following one another, enlarging the tear size faster and faster.
And so on, until when
EVERYTHING IS SUDDENLY TORN APART.

The canoe floated from a calm river into turbulent rapids, before being finally drawn into the FINAL FALL !

Nature is tactfully warning us!
But you have to learn how to listen!

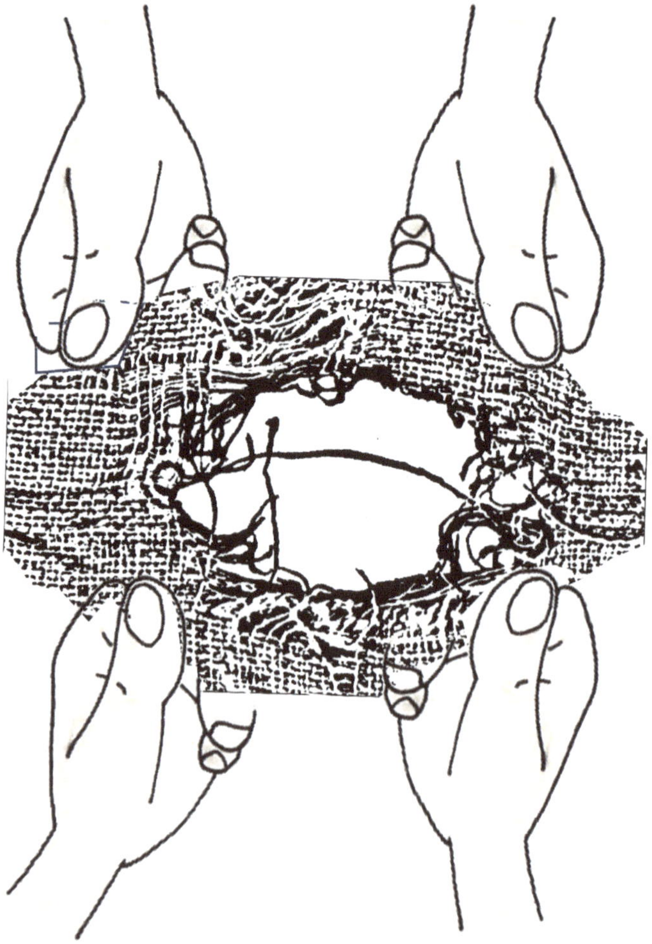

4.2 What Science predicts and what Nature is telling us

It's not doubt that drives you crazy, it's certainty (Nietzsche).

You were talking about the final plunge,
Eifel! Is it possible to predict the day and
time ??? Have we tried it yet ?

Yes, about fifty years ago !
We can't say we didn't know about it !
So, let's start with the precursors: Dennis and Donella Meadows, Jorgen Randers and William Behrens, in the report published by the MIT (Massachusetts Institute of Technology) at the request of the Club of Rome (1972).
In 1974, on this basis, agronomist René Dumont warned on the risk of a total collapse of our civilization in the 21st century (Dumont R., 1974).

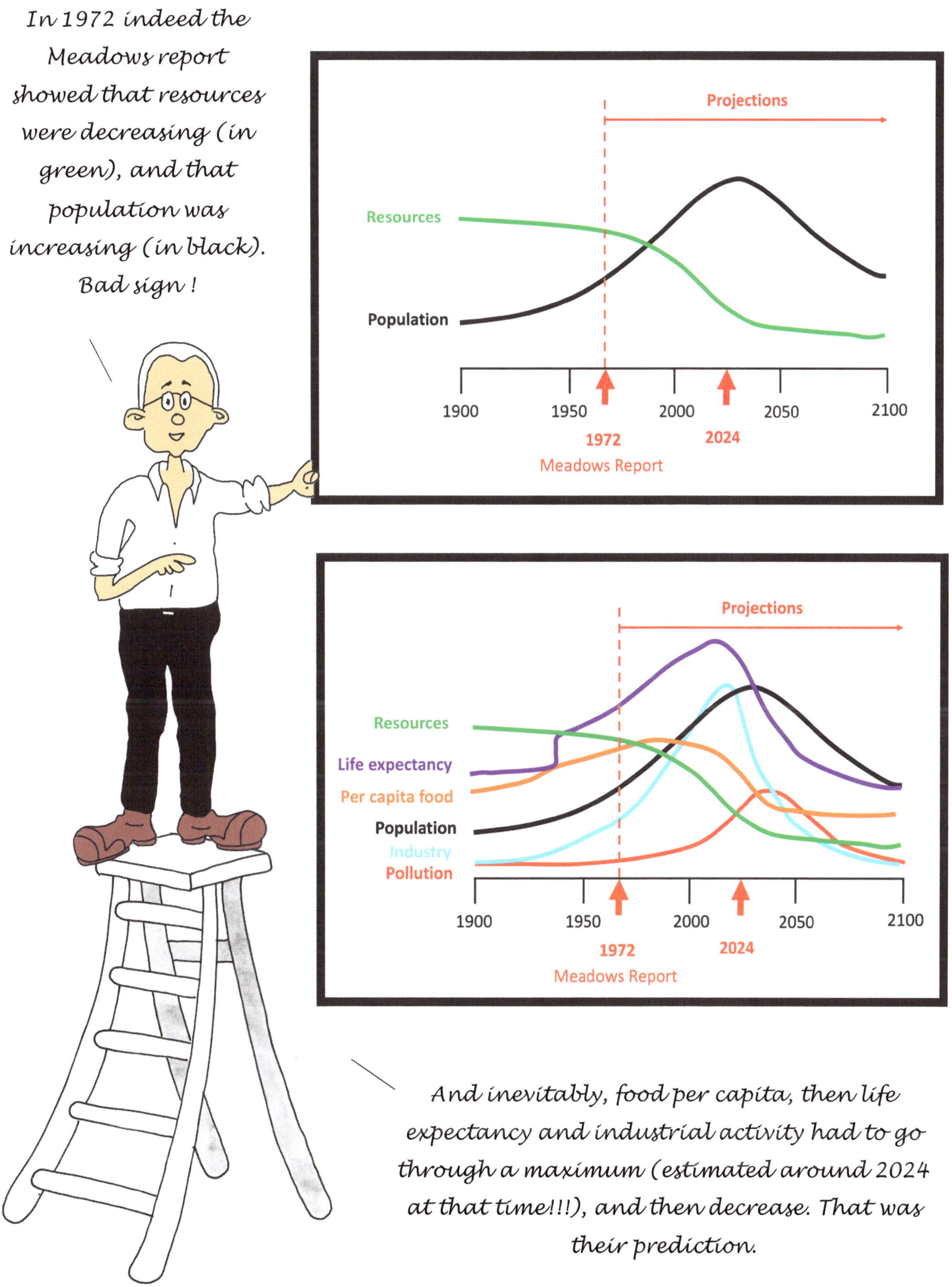
In 1972 indeed the Meadows report showed that resources were decreasing (in green), and that population was increasing (in black). Bad sign !
Projections
Resources
Population
1900
1950
2000
2050
2100
1972
2024
Meadows Report
Projections
Resources
Life expectancy
Per capita food
Population
Industry
Pollution
1900
1950
2000
2050
2100
1972
2024
Meadows Report
And inevitably, food per capita, then life expectancy and industrial activity had to go through a maximum (estimated around 2024 at that time!!!), and then decrease. That was their prediction.

2024 ! Well done! I've heard it's alas already off to a good start in several places on the planet
Impressive!

Wow! The Meadows report was quite impactful at that time indeed! And now, where do we stand???
Has there been any progress in forecasting techniques ?

Pablo Servigne and Raphaël Stevens had already spoken of collapse in their "little manual of collapsology" (Servigne & Stevens 2015). They explained that the various crises already experienced by humankind ("unexplained" vanishing of civilizations for instance) could herald a collapse of our industrial society (and its associated civilization), which would be global this time, on a planetary scale.

As for forecasting techniques, sure, they have improved in two areas. First, the spectacular computing power we got at the beginning of the 21st century, and second, the powerful applications of theoretical physics analytical tools derived from Henri Poincaré's famous work on deterministic chaos at the beginning of the 20th century (Holmes 1990). The theory of dynamical systems is one of them .

Henri Poincaré (1854 – 1912)
French mathematician, theoretical physicist and philosopher of science

There are therefore two main forecasting methods:

"Massive" numerical simulations ...

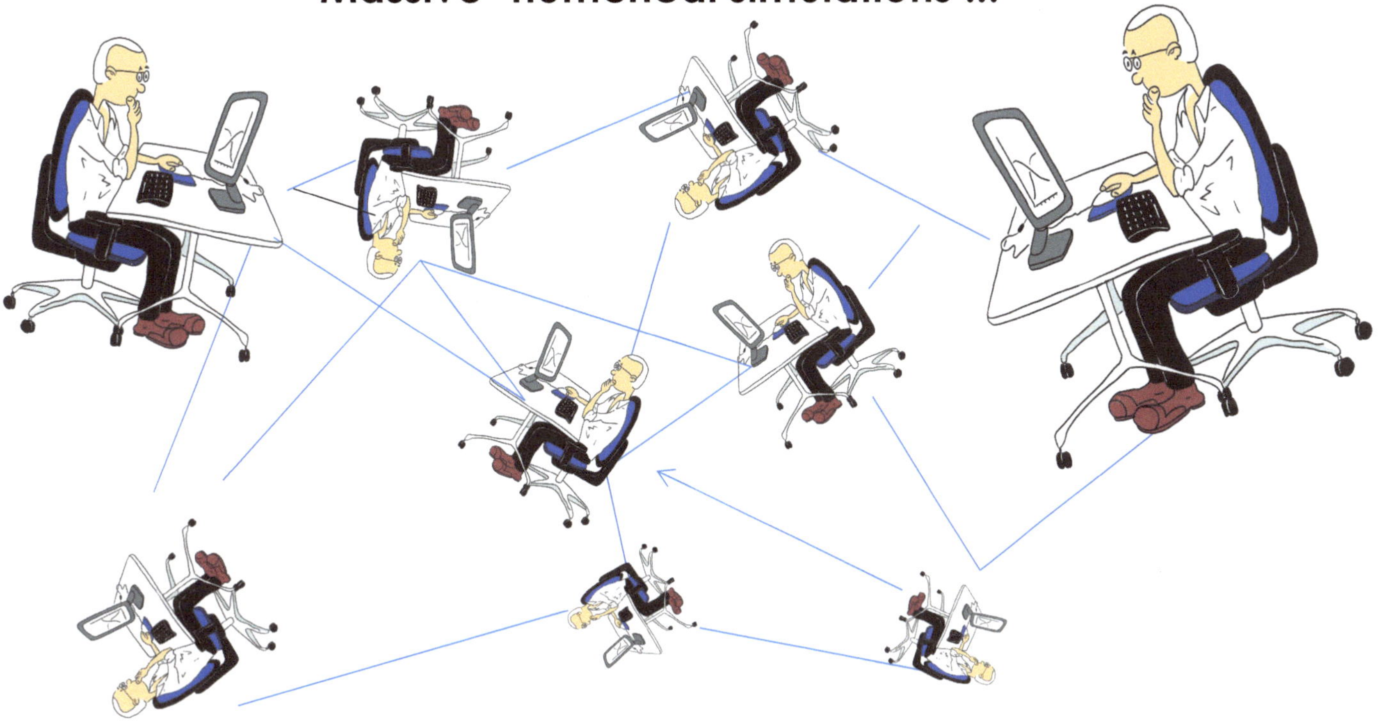

... and the theory of dynamical systems:

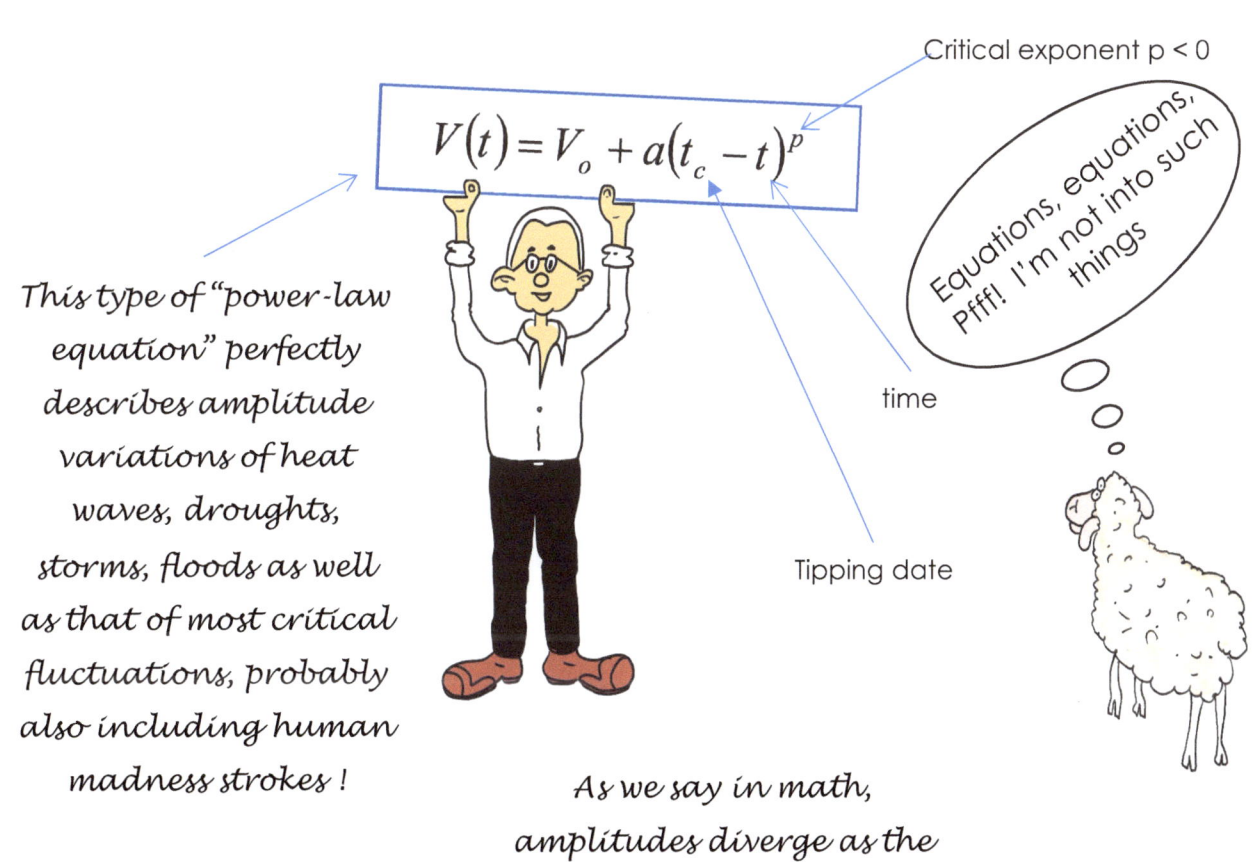

$$V(t) = V_o + a(t_c - t)^p$$

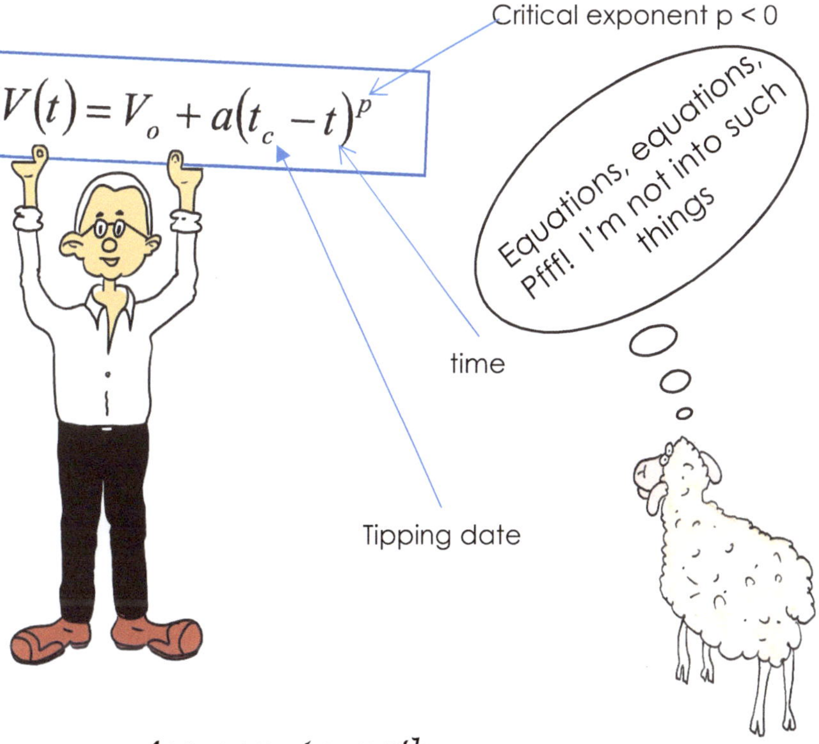

Critical exponent p < 0
time
Tipping date
Equations, equations, Pfff! I'm not into such things

This type of "power-law equation" perfectly describes amplitude variations of heat waves, droughts, storms, floods as well as that of most critical fluctuations, probably also including human madness strokes !

As we say in math, amplitudes diverge as the tipping point is approached !

4.2.1
Numerical simulations

It is characteristic of an educated mind to be satisfied with the degree of precision which the nature of the subject permits,
and not to seek greater accuracy
where only an approximation of truth is possible.
(Aristotle)

Now let's look at these methods in detail, starting with simulations.

The "ocean-atmosphere system" is divided into a large number of boxes, just like camera pixels.

Each of these cells is assigned initial conditions of temperature, pressure, humidity, wind speed, etc.

Then, these cells are asked to exchange such information with each other and respond by changing their own state, using laws of physics.

Some of these data include different possible scenarios of greenhouse gas emissions or political or economic decisions (or indecisionsetc), resulting in different forecast trends depending on the chosen scenario.

And calculations can be launched !

Simulation results can be found on the web, particularly on the IPCC (Intergovernmental Panel on Climate Change) website.

Below is an example of different scenarios where greenhouse gases are injected into the atmosphere

It can be seen that in ALL simulations, Earth's surface temperature change is CONTINUOUS (no sudden jump), from the most optimistic scenario (bottom) to the most pessimistic one (top).

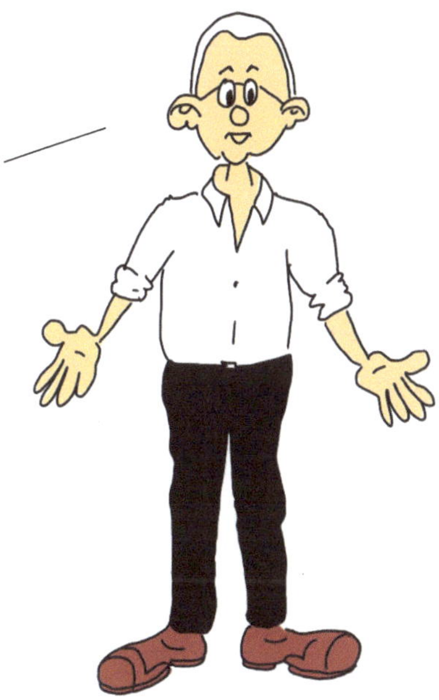

Typical results of the different greenhouse gases injection scenarios into Earth's atmosphere

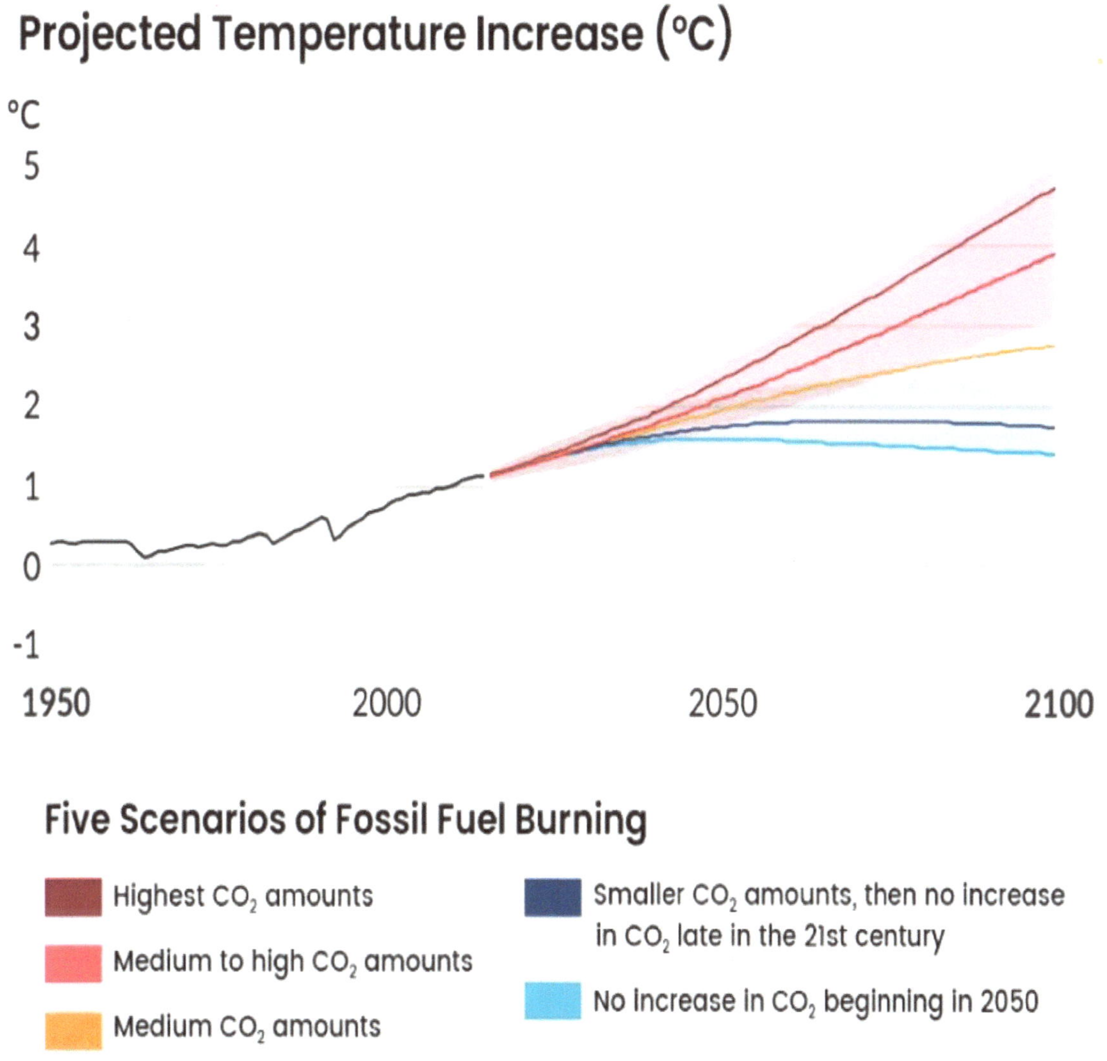

From: https://scied.ucar.edu/learning-zone/climate-change-impacts/predictions-future-global-climate

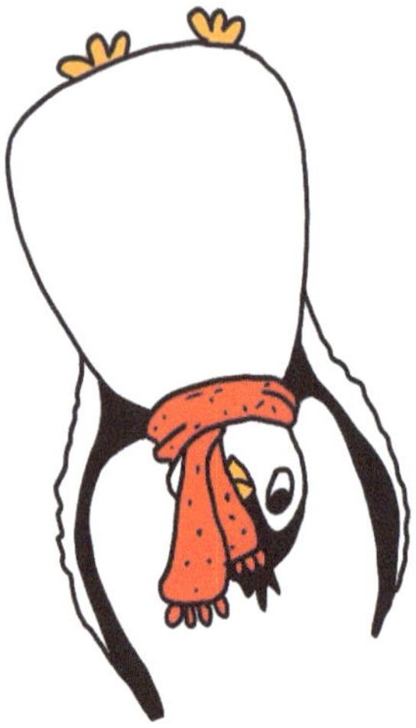

What a machinery! It's mind-blowing !
But I've heard that these forecasts have to be recurrently and systematically revised upwards. How is that possible ?

Oddly enough, yes, it's always upwards! This has led some people to postulate a supposed evolution of a so-called "climate sensitivity" to CO_2 addition
(Sherwood et al., 2020, Saint-Martin et al., 2021)
in order to "explain" such upward adjustments.

Obviously, this is only a tinkering that cannot explain anything!

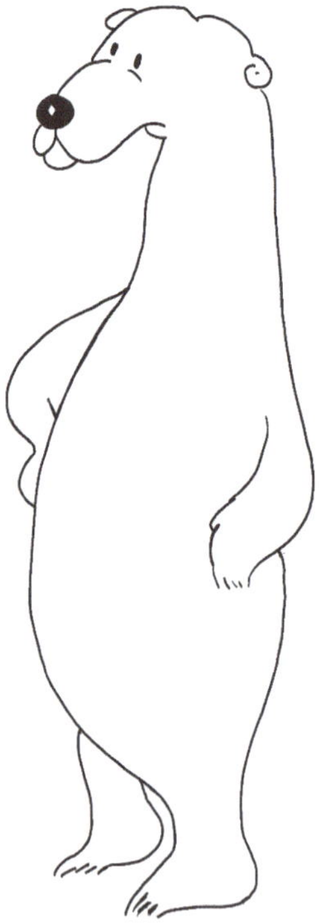

But is warming really continuous ? Can we make sure that we will escape the crack destabilization in the burlap, or Churchill River cataract ?

Not sure at all !
To get a clearer picture, let's get back to Physics !

4.2.2
Tipping points and warning signals

"It's a sad thing to see that nature speaks
and that mankind does not listen to it."
(Victor Hugo)

Whoever picks a flower disturbs a star
(Théodore Monod)

I'm a simple being, the dog told me !

But the shepherd also told me that when Barky herds us, we become very difficult to manage."

On my own, I'm simple.
Together we are complex !

Extrapolating the behavior of an isolated sheep ...

... cannot predict herd behavior.

As physicists say, such a feature is called "emergence of a global macroscopic behavior, absent at the scale of individual elements"

This interesting property is typical of so-called "complex systems", that are made up of a large number of strongly interacting elements, like a sheep flock that has been herded.

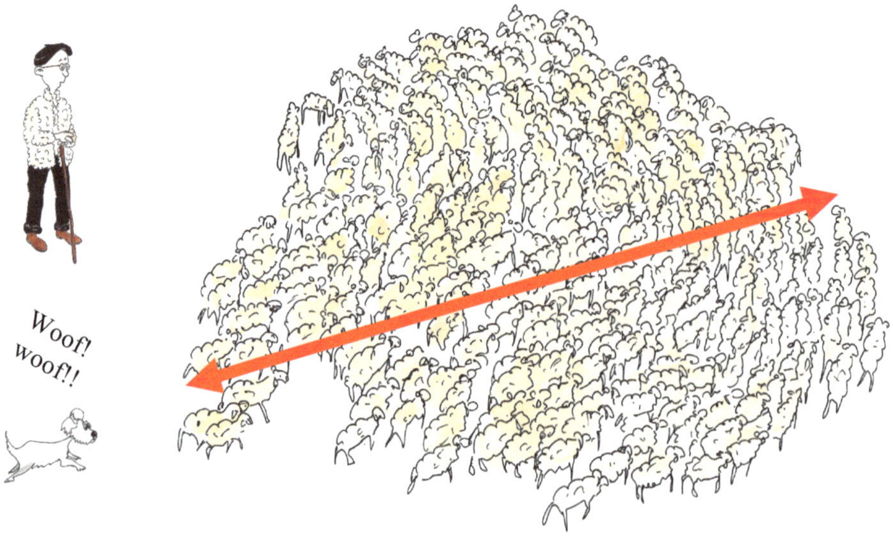

When the flock is dispersed, the behavior of a single sheep affects only its nearest neighbors, within a distance of a few meters, which is called the "correlation length".

But as the dog gathers the flock, the sheep jostle each other more easily, and the correlation length increases.

Then comes a time when it reaches the size of the herd (in red on the figure).

At this stage, the behavior of a single sheep can affect the entire flock.

Is this what is called "critical point" ?

Yes, exactly.

This is where a tipping point can be met, such as when one of the sheep "accidentally" decides to jump a cliff, dragging the whole flock with it.

So, I'm spending my time making critical points without knowing it ???

Yes, of course, and that's your job, Barky!
I don't blame you for anything !
I can't let the herd disperse; that would make
the wolf too happy.
If the flock jumps off the cliff, it's not your fault,
it's the butterfly's fault.

Butterfly? Which butterfly ?

Imagine a butterfly delicately landing on the snout of a sheep. This sheep, which was about to jostle its neighbor on the right, will jump up and jostle the one on the left. This will completely change the shape and extent of the jostling cascade that will follow. Instead of jumping the cliff, the herd may possibly end up on its way to the sheepfold...

or (but this is less likely) cause
a tornado on the other side of the planet.

That's not a good sign!

The butterfly effect again!!

What? Myrtle??

Why are we changing direction?

She saw the butterfly!

We're going to jump into the ravine!

I'd prefer return to the sheepfold.

It's Myrtle

Depends on Myrtle!

I would rather say on the Butterfly!

And the butterfly doesn't obey Barky!

If the shepherd can't manage the herd he has gathered, would his computer be able to handle it? And what about the people managing the climate simulation computers?
Interesting question! Just like the cells of numerical calculation systems, sheep interact with one another. If the flock is not too big, the shepherd's computer may easily calculate all the possible cascades of events due to sheep interactions.
You're looking thoughtful, Barky...

But compared to a flock of sheep, the number of cells involved in climate calculations is huge. And, even worse, the number of interactions between all these cells grows astronomically, especially when we approach the critical point where any sheep may trigger a series of events on the scale of the entire flock.

The computer network may then saturate and lock up. In such situations, people usually resort to "simplification procedures" .
For instance, it may be decided to divide a flock of 100 sheep into 10 groups of 10 sheep each, which will be called "super sheep", and to assign each of these super sheep the average properties of the sheep they are made of.
It's easy to understand that the number of interactions between these super sheep will be far less than it was in the flock of real sheep.

The same argument shows that it is easier to simulate the behavior of bathers on a beach during a heat wave by gathering them under umbrellas. It's indeed faster to manage interactions between a small number of umbrellas than between a large number of individual bathers.

And the calculation can start again!

The calculation can start again?
Are you kidding, Eifel?

In my small head, I understood that all these chains of fiber ruptures resulting in fabric failure, the jostling of sheep leading them to jump over the cliff, and all those turbulent shocks announcing and producing the final climate tipping were likely to occur only for a gigantic number of interactions! Right ???

Well done, Seagull! You're amazing !
In any case, it shows that intelligence is not directly related to brain size !
And that you can observe things from far above.

Yes, indeed. This "explosion" in the number of interactions in the vicinity of the critical point is a characteristic (if not a definition) of this critical point.
The numerical model will give an acceptable description of the "super sheep" flock behavior as long as it remains far from any tipping point.
It will never be able to see them leaping off a cliff!

And in the same way, numerical calculations can simulate climate change as long as the REAL climate is not heading towards a tipping point in the short term.

But if this was the case, the super sheep wouldn't realize what's really happening... And forecasts should be revised upwards more and more frequently, without being able to catch the slightest glimpse on the incoming crash, as in the case of the boater who quietly bails out his canoe while hurtling towards the cataract.

That's it, that's it!
That's what I was saying earlier!
If you can't predict anything, you're really up to your beak in guano, as we say at home!

Well, it's okay, Seagull, I don't know how to chain loops and tight turns, and laugh like you, but let's be positive.
If simulations don't work in the case of tipping, there may be something else to do ???

And first of all, do all of these tipping points really exist? Why would we be concerned ?

Yes, we know they exist. Simply because paleoclimatologists have analyzed similar climate tippings that have actually taken place in the past. And some of them have considered a possible application to the current situation (Livina et al., 2007, Lenton 2011, Lenton et al., 2012a, 2012b, Bathiany et al., 2016, Steffen et al., 2018, Lenton et al. 2019). And we are already seeing warning signs of this.

In a word, Penguin, yes., there is something to do . In fact, I think it's high time to get back to the question at hand. Instead of trying to predict Nature's behavior through simulations, let's try to listen directly to Nature itself, to understand what it is telling us right now.

Wouldn't you by chance allude to the jolts that announce the approach of a tipping point ?

You've got it all figured out, Penguin. In physics, such jolts are called "precritical fluctuations", or "critical softening".

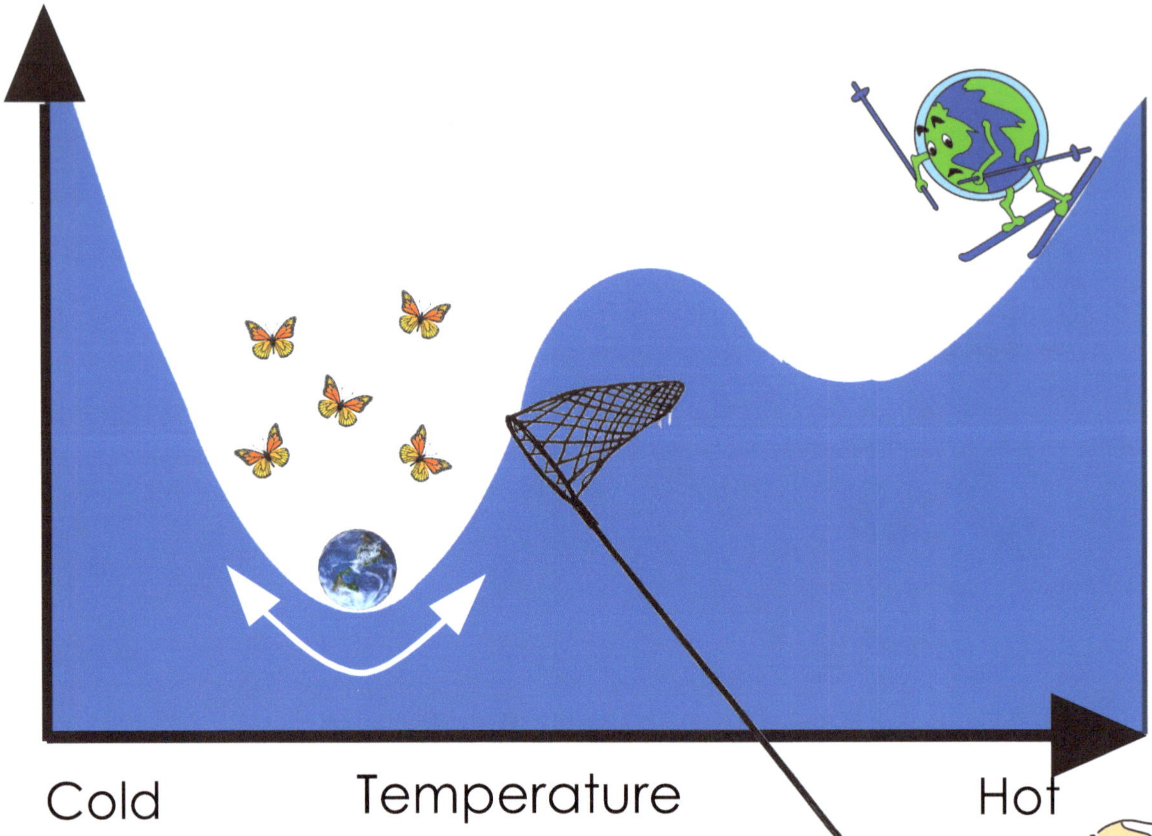

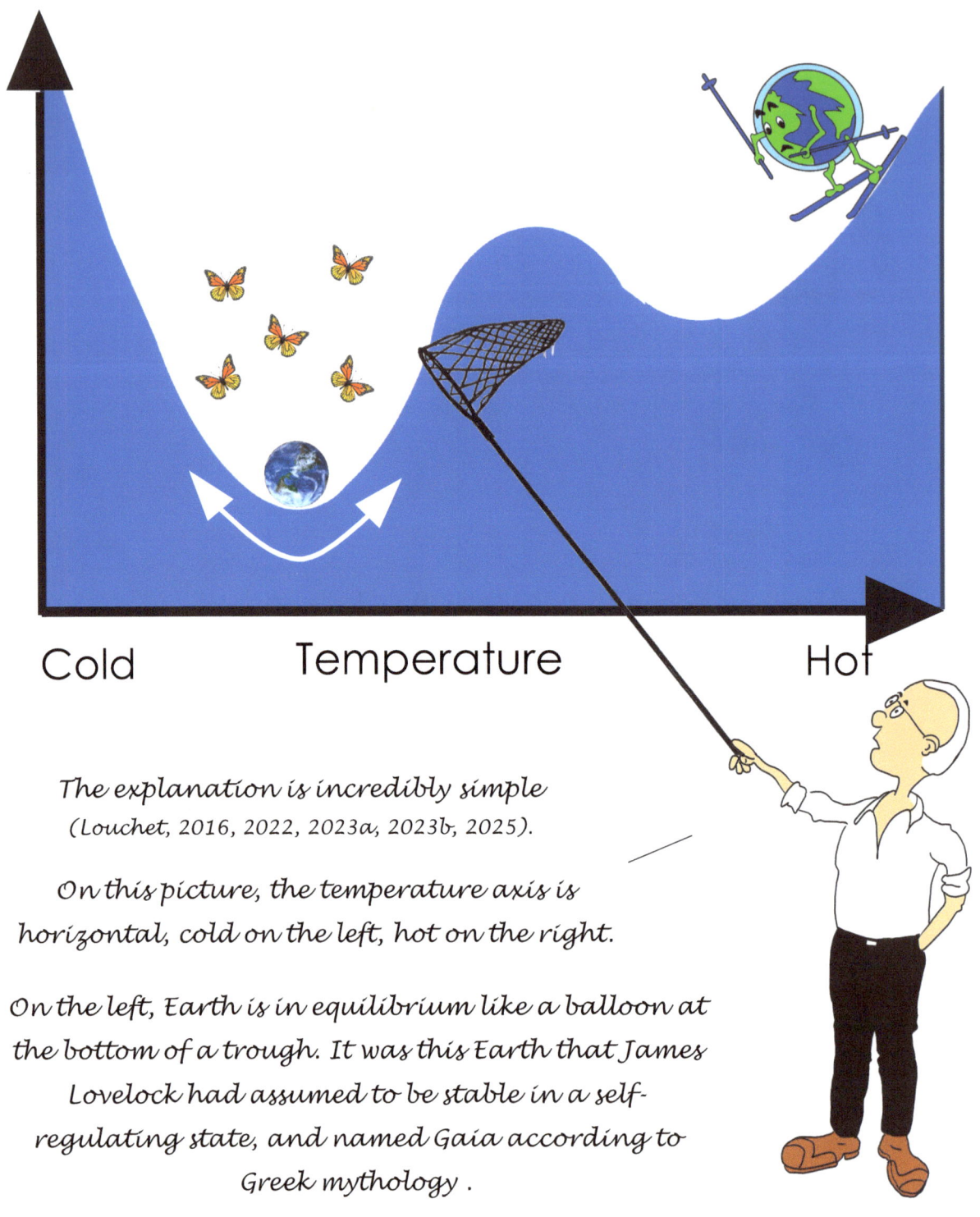

The explanation is incredibly simple
(Louchet, 2016, 2022, 2023a, 2023b, 2025).

On this picture, the temperature axis is horizontal, cold on the left, hot on the right.

On the left, Earth is in equilibrium like a balloon at the bottom of a trough. It was this Earth that James Lovelock had assumed to be stable in a self-regulating state, and named Gaia according to Greek mythology .

Because of a few unconscious butterflies flapping their wings in New Zealand or elsewhere, the balloon representing the Earth swings slightly forth and back, like so many temperature fluctuations. These are the usual small heat strokes or cold snaps, around an average temperature value close to the valley bottom.

But if we now gradually increase the amount of CO_2 in the atmosphere, the profiles of these valleys would be distorted. Earth would still stay in equilibrium at the bottom of the trough, but the trough itself would have shifted to the right. "Average climate" would have turned warmer. In addition, as the valley floor flattens, disturbances coming from antipodes or elsewhere would result in increasingly important oscillations of temperatures, winds, etc, resulting in a broadening of oscillation distributions. Heat waves and cold snaps would turn stronger and stronger, and so would rains, droughts, storms and associated forest fires.

In addition, and this is a key point, the valley flattens more on the right side than on the left one, due to the proximity of the tipping point. The distribution becomes more and more asymmetrical. with an increasing weight of largest events. All these observations are clearly illustrated in:

https://svs.gsfc.nasa.gov/5452

These are the famous precritical oscillations, known in climatology as extreme events! The relevance of average values turns then untrustable, due indeed to the disproportionate role of rare but major events*. This growing asymmetry shifts the average value beyond the bottom of the valley profile, that might hide the possibility that such rare but large events may exist. This is called the Black Swan theory* (e.g. Taleb 2007, 2010).

* https://en.wikipedia.org/wiki/Black_swan_theory

Not surprising! It was a common belief tin Europe that black swans didn't exist at all, until Dutch mariners discovered some of them in Australia in 1697.
But let's go back to extreme events. In the vicinity of a critical point (tipping point in our case), data obey so-called power-law (also named scale invariant) statistical distributions. https://en.wikipedia.org/wiki/Power_law.critical
They are quite different from bell-shaped ones, in that they are not symmetrical, but exhibit "fat tails", accounting indeed for scarce but unusually large events, whose rarity may make them escape our vigilance. Averaging several successive computation results in order to get rid of the butterfly effect, as reported for instance by (Lecroart & Ekeland 2020, p. 33) becomes here questionable. When approaching a critical point, we should remember Gerald Durrell's sentence: "As the buoys marking the shoals are often out of position, mariners are cautioned to be on their guards when navigating these zones" (Durrell 1959).

Such precritical oscillations, evidenced on the example of temperatures in: https://svs.gsfc.nasa.gov/5452 as mentioned above, can be clearly illustrated carrying out a small experiment with a cork on a cardboard, which reproduces the evolution of the previous diagram, as shown in the following video:

www:https://www.youtube.com/watch?v=EN16HcawPXU

No, Eifel, you can't make me believe that! I still remember you saying that the climate is a complex system made up of a huge number of interacting elements! It's a very complicated thing. And you think you're going to convince me with a simple cork on a vulgar piece of paper?

And yet I do! This is exactly what is shown by records of land temperatures measured since 1964 and recently published (2025) by NASA in the beautiful video I mentioned above:

https://svs.gsfc.nasa.gov/5452

(These data are still available, but due to the lapse in federal government funding, NASA is not updating this website any more.)

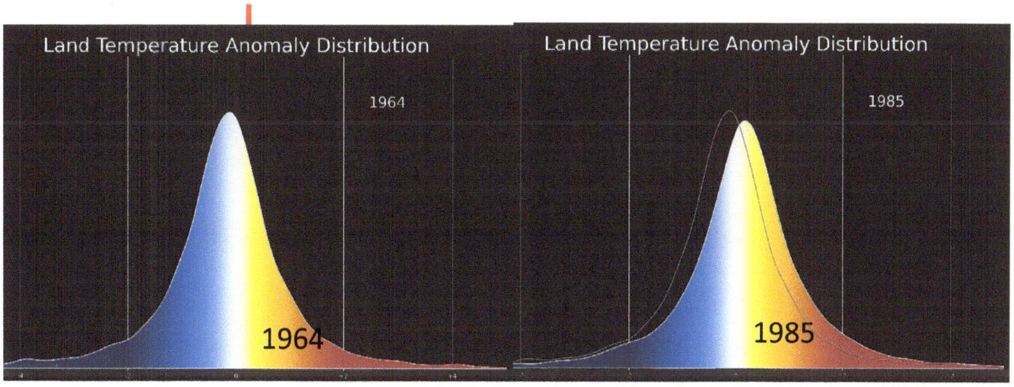

It shows that from the 60s to the 90s, the temperature distribution shifted to the right, but remained relatively tight (the starting position is shown as a thin line in the figure on the right). Temperatures fluctuate slightly and symmetrically around the mean value, but ......

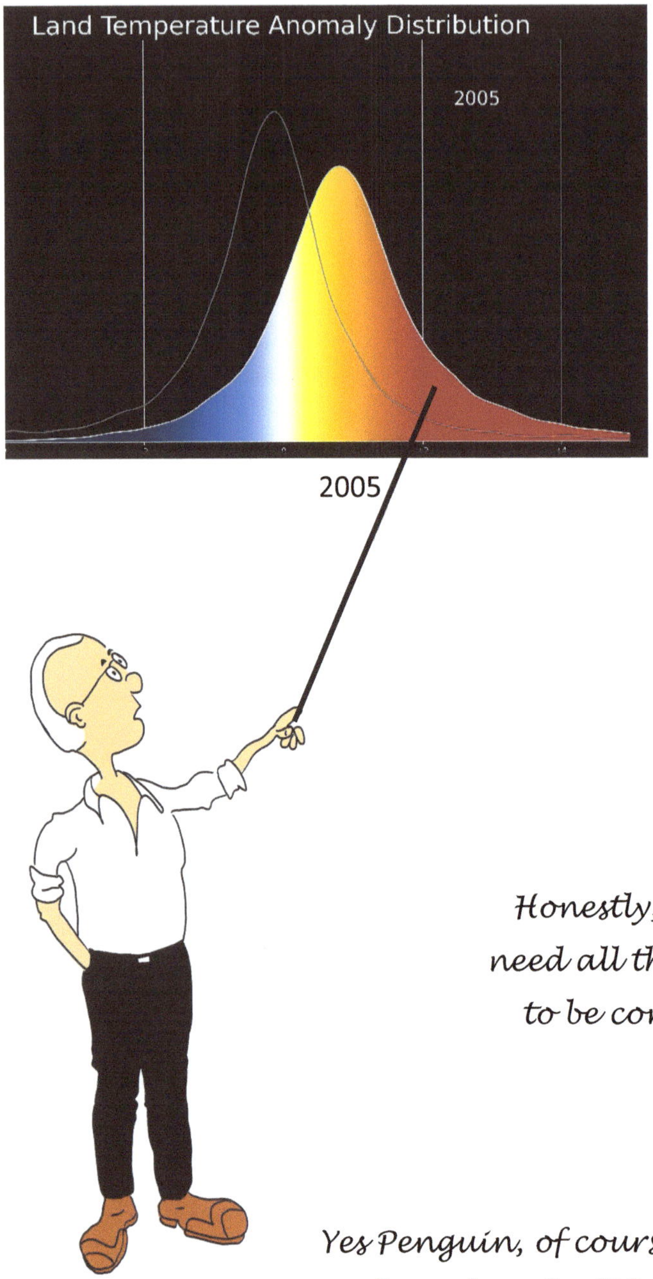

... however, in the following decades, things take a turn for the worse. As warming continues (the starting point is shown by thin line), the distribution begins to broaden as it shifts towards higher temperatures, meaning that the amplitudes of fluctuations increase. More intense heatwaves and cold snaps, obviously associated with more violent storms, more devastating floods and droughts, etc...

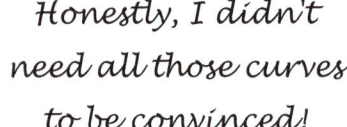

Honestly, I didn't need all those curves to be convinced!

Yes Penguin, of course, but Seagull really seemed to need it to admit that the "blue" diagrams on the previous pages and the cardboard and cork experiment made sense!
This NASA publication totally confirms what the famous blue diagrams and the cardboard experiment show, but also what we're feeling year after year. And in this publication, these measurements give us precise, quantitative and irrefutable proofs.

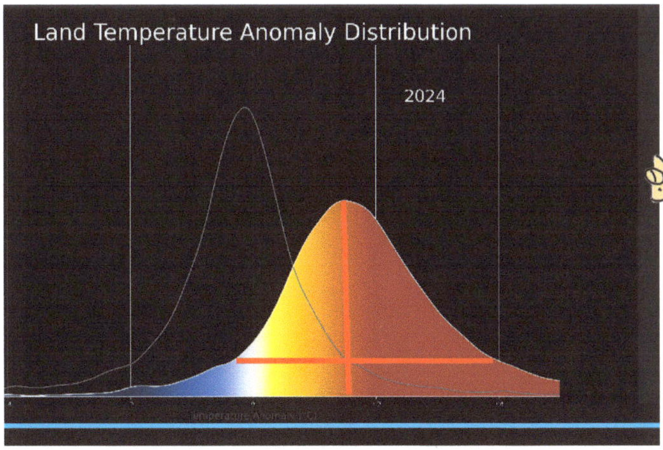

2024

But even more interesting: over time, the curves become increasingly asymmetrical. They widen more towards the right than towards the left of the distribution peak. This is exactly what the blue diagrams on the previous pages predicted.

*So what?
What the hell
does that mean?*

*Exactly what Eifel was saying right now!
It's one more piece of indisputable evidence
that we're approaching a tipping point.*

And now I'd like to insist (again) on two things. The first one is that this is a global effect, a new property that emerges from the interactions between all the elements of the system. The second one is that this distortion of the distributions to the right is what we qualitatively expect to see in the gradual transition from a "Gaussian" bell curve (low-correlation events) to a power law, characteristic of a highly correlated system (critical state), in which the most probable events are the smallest ones, while the largest ones are the least frequent, the famous "black swans" responsible for the "thickening" of the tail of the distribution on the right. Similar results are found for example in Bak's sand pile model defining the notion of self-organized criticality (Bak et al., 1987).

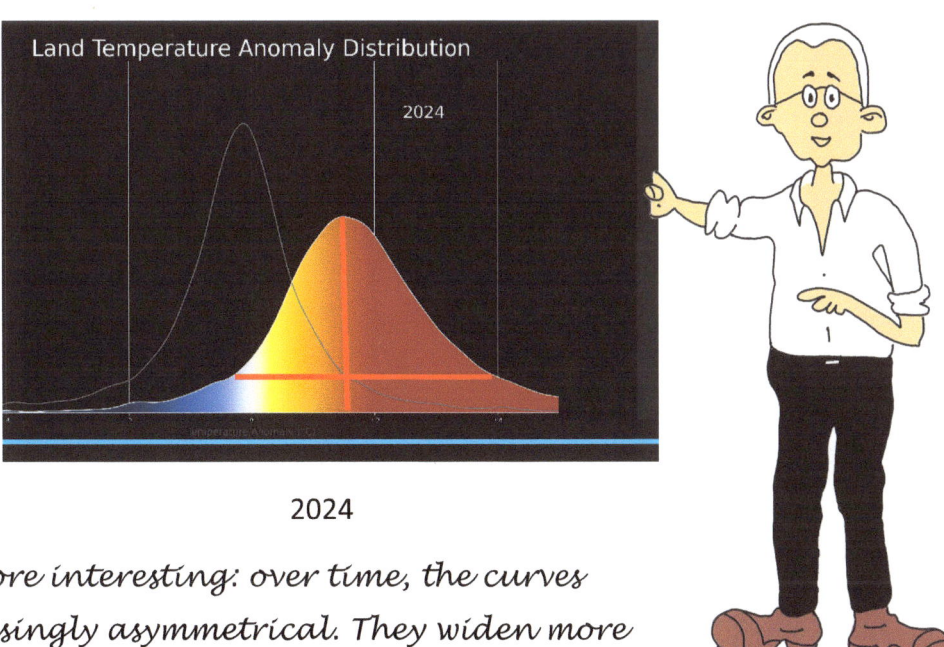

But what, if we decided to reduce now the amount of CO_2?

Yes of course! Obviously, we would go back to the starting position.
But we would have to decide and do it very, very quickly, otherwise, if we delay too long ...

... our poor planet would suddenly fall over into the other side, to the bottom of a nearby valley, about which we know absolutely nothing, since no one has ever been there before. Nothing, except that it would be significantly hotter, and for long.

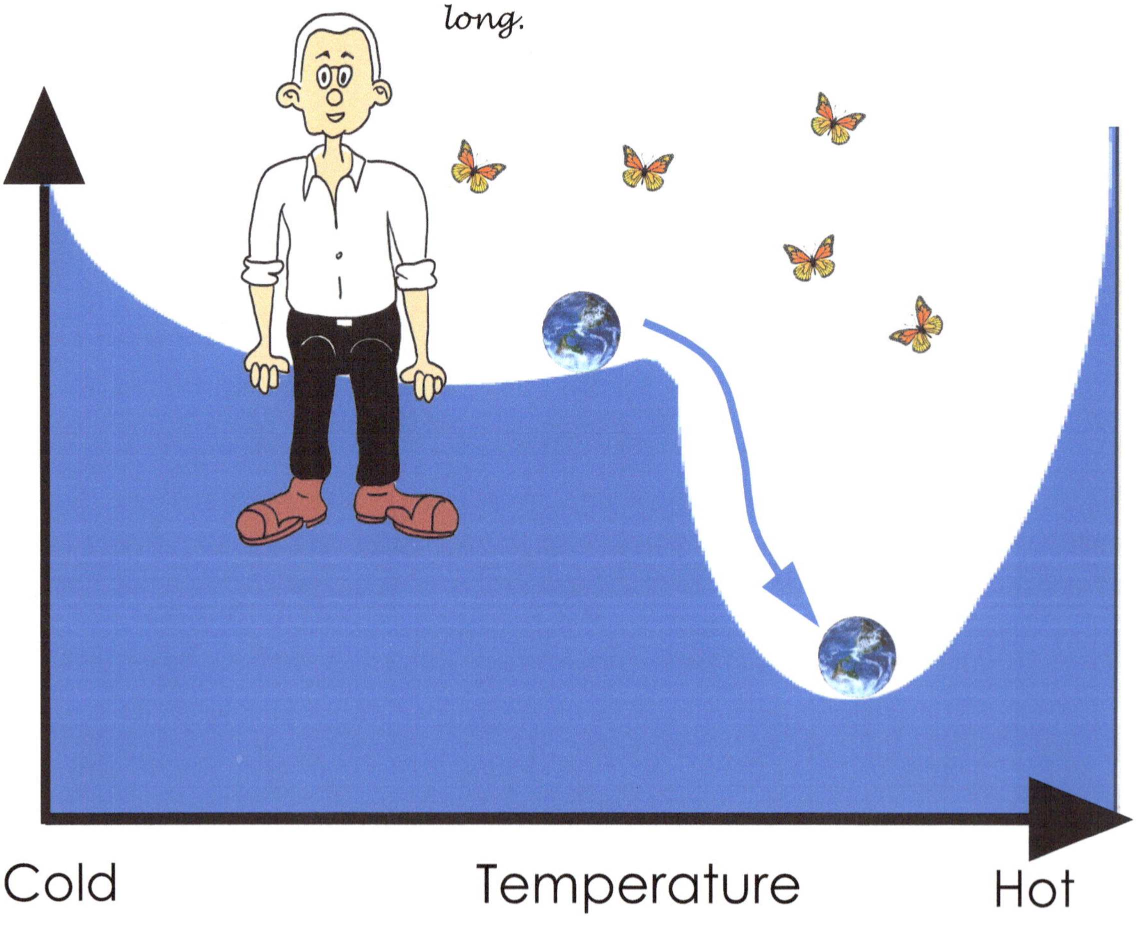

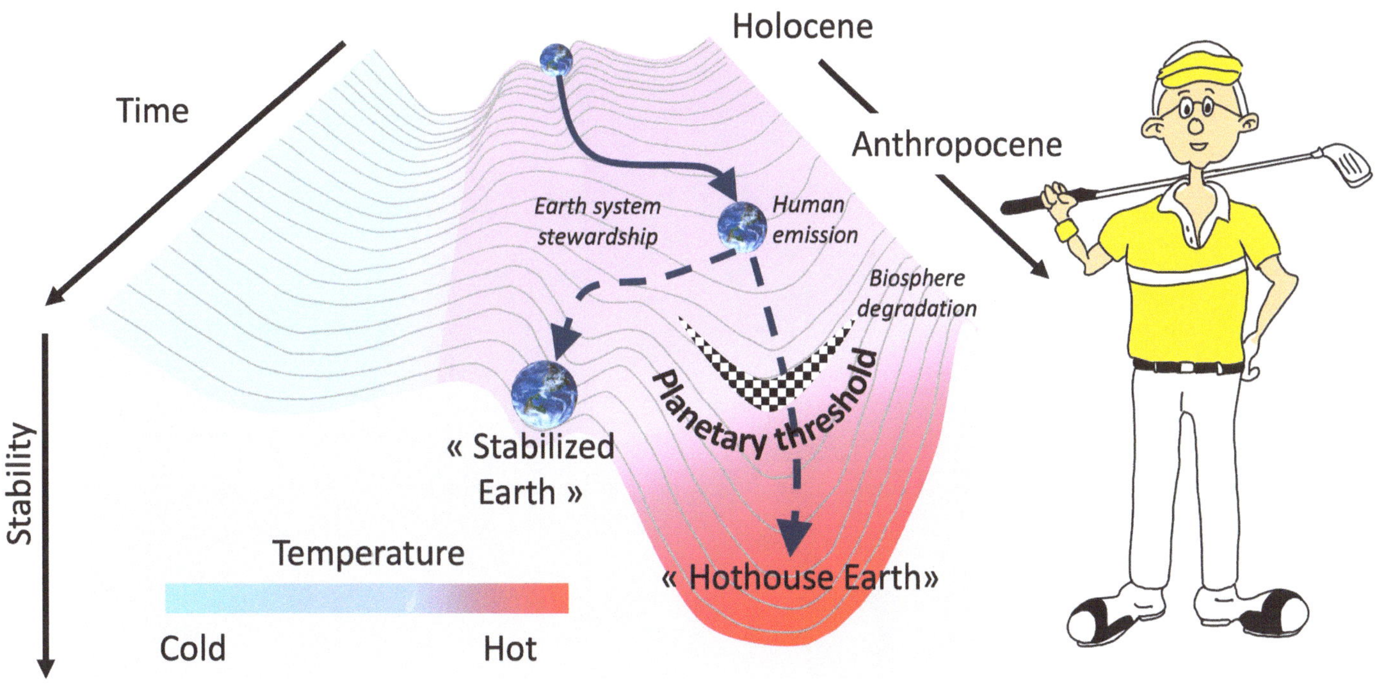

From: Steffen, W. et al., 2018

This process can be illustrated in another way. In this diagram, time flows from the bottom of the diagram (Holocene) to the front of it (Anthropocene), and "stability" is represented on the vertical axis (the lower, the more stable). As for temperature, it is illustrated by the color, cold in blue, warm in red.

Our planet comes "from the depths of time", and rolls forward, following the bottom of the valley. But at some point, the ridge separating its valley from the warmer one on our right gradually fades away. Earth hesitates between staying in the initial trajectory (blueish) and shifting towards the hotter one (in red). The slightest perturbation would be enough to steer it one way or the other.

That's where we are. Will we be able to bring the little piece of wisdom that could allow it to stay in the blue, or will we let it branch off into the red ?

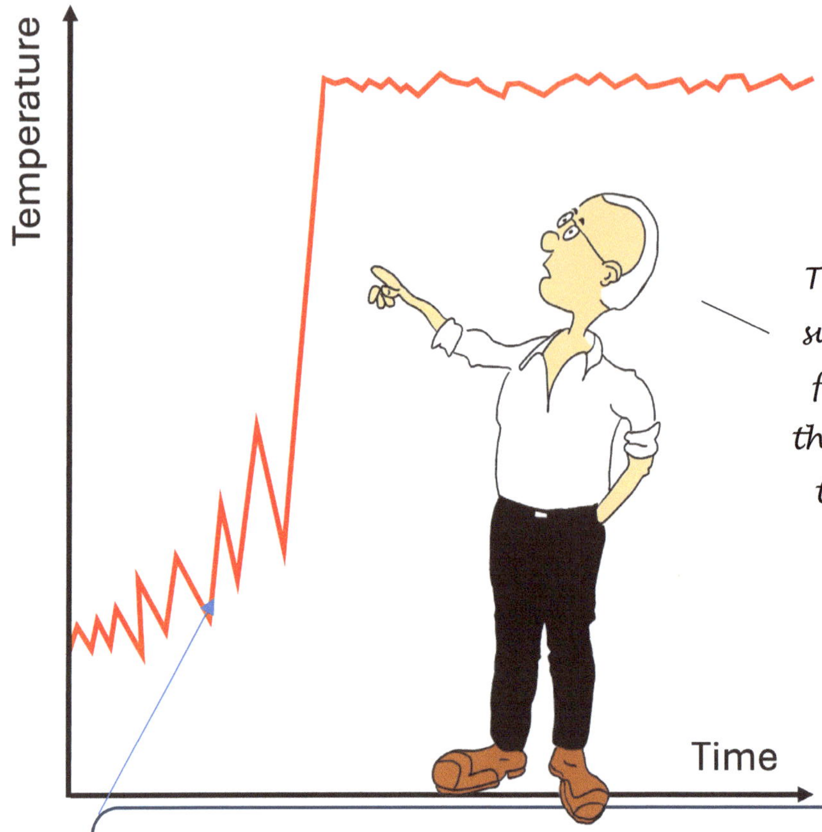

This jump would correspond to a sudden increase in temperature, from that of the saddle point to that of the bottom of the valley on the right of the previous figure.

Succession of extreme events warning of an incoming tipping point

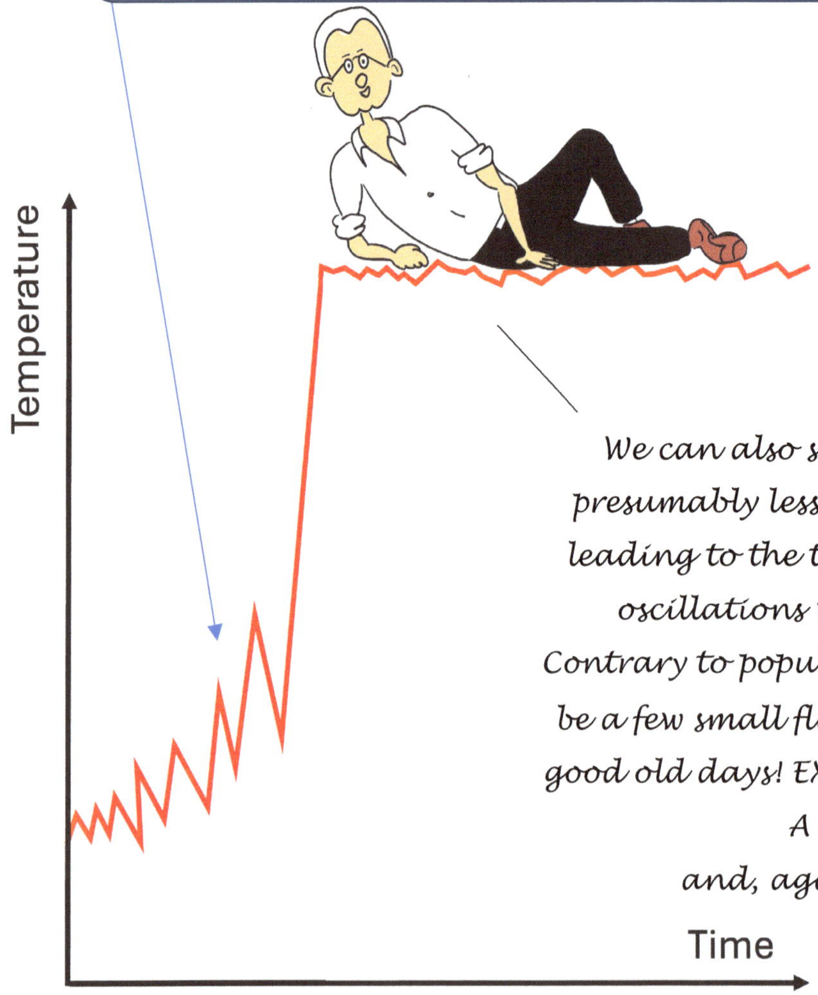

We can also see that in this new valley, presumably less flattened than the plateau leading to the tipping point, the precritical oscillations would have disappeared. Contrary to popular opinion, there would only be a few small fluctuations left... Just like the good old days! EXCEPT THAT IT'S GOING TO BE A LOT HOTTER...

and, again, for a long time!

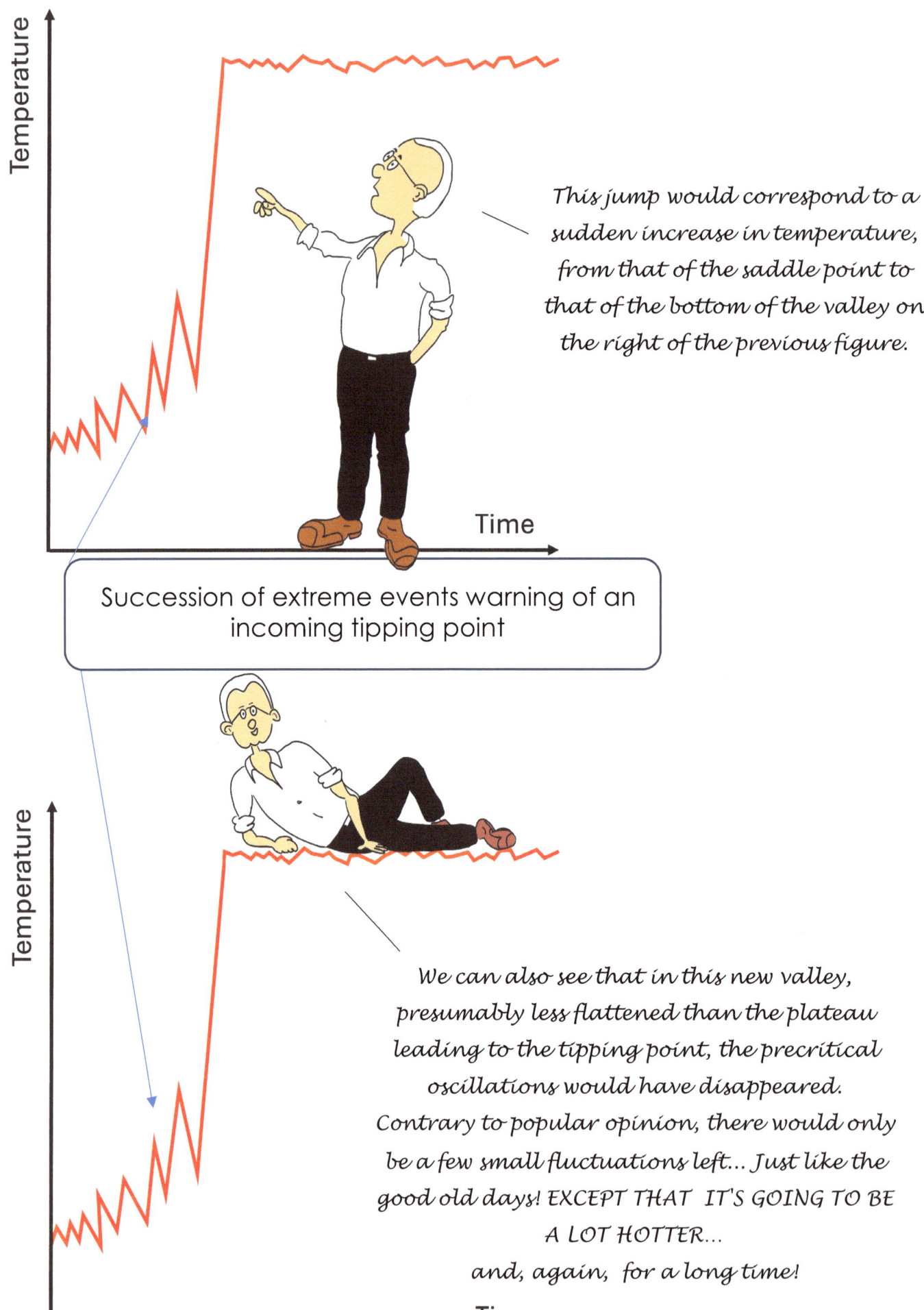

Based on a thorough analysis of these precursory signs, a team of geophysicists from Zürich was able to develop a method a few years ago which predicts the date of serac collapse in glaciers (Faillettaz et al., 2015, 2016). They recorded fluctuations in glacier flow velocity by means of GPS beacons. The "runaway" of these warning signs allowed them to determine the date of the collapse with an IMPRESSIVE accuracy, from 24 to 48 hours, which made it possible to evacuate in time the villages located downstream.

An analysis of the same type, based on the approach of a critical point, was also used at Copenhagen University to predict the time slot for a collapse of ocean currents circulation in the Atlantic, or "AMOC" (Peter & Susanne Ditlevsen 2023), which is a key element in the climate bowling game, as we will see later on.

So, if I understand you correctly, Eifel, we won't have any more heat waves, cold snaps, storms or floods in the post-tipping period, but it will be much hotter, right ?

Er, just a minute! I'd be inclined to agree with you, Penguin. But I also often hear it claimed that these episodes of intense rainfall we're experiencing, causing flooding and other disasters, are due solely to the rise in temperatures caused by global warming. A warmer atmosphere would contain more water vapor, and therefore produce more rain.

Seems true at first glance. The well-known Clausius-Clapeyron equation represented by this graph (top curve labelled 100%) shows indeed that the maximum amount of vapor the atmosphere can hold <u>at a temperature of around 20°C</u> increases by about 7% per additional degree. It would be obviously more. at higher temperatures,

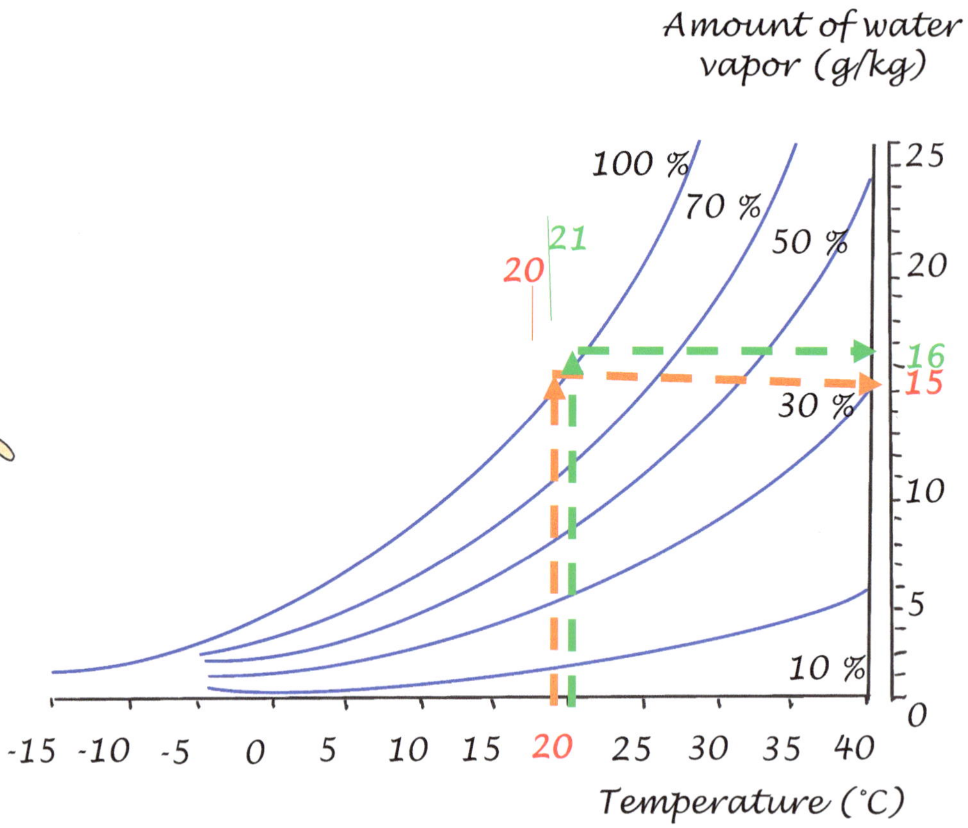

Variation of saturation water pressure with temperature (top curve) for an atmospheric pressure P= 101.325 kPa. Lower curves stand for unsaturated vapor.

From: https://www.eoas.ubc.ca/books/Practical_Meteorology/prmet/Ch04-Moist.pdf

Hey, 7% is something, isn't it?

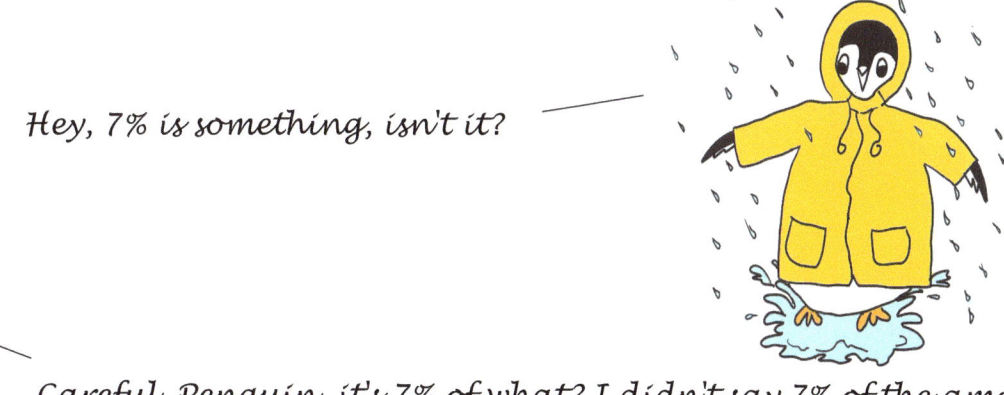

Careful, Penguin, it's 7% of what? I didn't say 7% of the amount of vapor you'd have above a pot of boiling water! Let's just notice that water boils at 100°C because 100°C is the temperature at which the saturation vapor pressure becomes equal to usual atmospheric pressure, which can no longer "hold it back". What I'm talking about, on the other hand, is 7% of the quantity of vapor present in air already saturated at 20°C before the temperature is raised by 1°C. You can see from the diagram that around 20°C this "saturating" quantity of vapor is 15g / kg of air, i.e. 1.5%. And 7% of 1.5% is roughly 1 / 1000!

Ha ha ha! Ridiculous! But then, Eifel, I can't understand how to explain all those flooded fields, overflowing rivers, that I can see flying over many countries!

My dear Seagull, in your travels have you even seen the Sahara or the Central Asia deserts turning green? Suppose for a while that these 7% are the right explanation. How then are we to understand the dreaded droughts that alternate, depending on the place and date, with these downpours?

Moreover, the graph below shows that there was not the slightest sign of an increase in the world average level of precipitation since ... 1960, while the average temperature rise since that time has already exceeded 1.5°C.

As you would say, Eifel, when observations disagree with theory, it's usually not the observations that are wrong!

That's right!

The explanation obviously lies elsewhere.

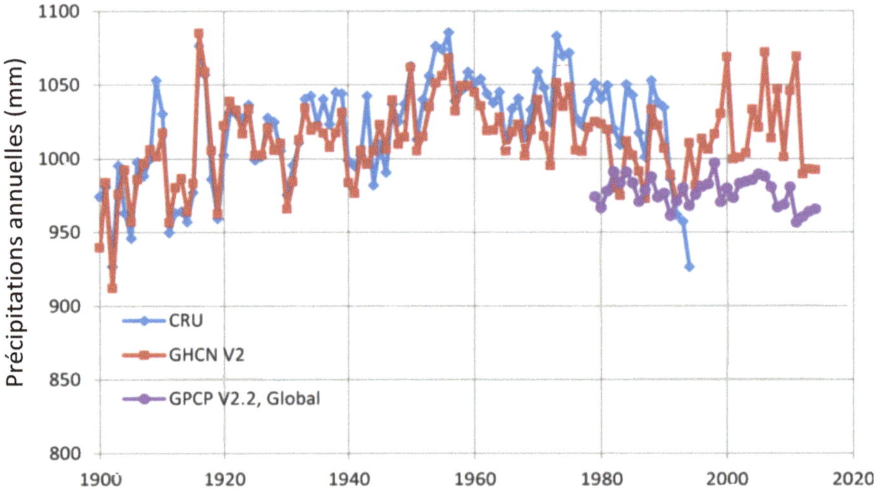

Evolution of average global precipitations after various data banks.

http://climexp.knmi.nl/select.cgi?id=someone@somewhere&field=gpcp_22), http://www.cru.uea.ac.uk/cru/data/precip/, (http://www.ncdc.noaa.gov/temp-and-precip/ghcn-gridded-products/).

Ooh, I presume you're thinking (again) about those pre-critical oscillations you were telling us about a moment ago?

Definitely. It seems to me the only reasonable explanation. It's the result of local fluctuations in time and space. The atmosphere is not always and everywhere saturated with water vapor, often a long way off... And it only rains somewhere when vapor reaches saturation locally, due to a temperature fluctuation, or to an incoming humid air mass in a cold zone, for example. And so it's not because the maximum allowed amount of vapor would increase by 7% that there would necessarily be 7% more rain!!!! In average, it's not raining more, as shown on the above curve, but it is raining more violently and more intermittently. And there will be times when ...

... Not a drop of rain, not a breath of wind, I'm dreaming of the storms of yesteryear !

You say it will be much warmer, but it won't be a heat wave ???

No, it won't be a heat wave. A heat wave is a period that is very hot compared to the "normal" temperature and is followed by a return to normal. Whereas here it will be very hot all the time, and this will be the new "normal" temperature.

And couldn't there be heat waves on top of this already very hot period ?

Yes, that's not out of the question. But they would be "small heat waves", followed by moderate "cold" (or less hot) spells. Because the valley being presumably narrower, the oscillations would be weaker.

But again, Eifel, if we decided, in this scenario, to drastically reduce not CO_2 emission rate but total CO_2 atmospheric content back to pre-industrial levels, do you think we could go back to a milder climate?

You're talking nonsense, Penguin! For someone who can't fly, you should realize that to get back to normal, you'd have to climb up to the saddle point and cross the barrier towards the left. But at this stage there would be no driving force to push Earth back as you imagine, and our poor planet would remain stuck in the hot trough on the right side!

Right again, Seagull! We'd be stuck there. It would be a new Gaia, but fiercely hotter than the present one. We would probably have to wait a few hundred thousand years for an unlikely "alignment of planets" that would add to our possible efforts to reduce CO_2 amount and help us turn the clock back . Except that when I say "we", it's more than likely that we'll all have "evaporated" by then !

As for me, at present, I think that I will be forced to sell (but to whom?), and quickly, all my parkas, mittens and fur-lined hats, and build myself a straw hut in Greenland! If I choose my spot carefully enough, after ocean rising, I will find myself on a small desert island north of the Arctic Circle. During this time, Flanders would have largely disappeared, as well as Camargue or Pacific atolls, and Paris will no doubt be by the sea!! That would certainly reduce vacation traffic. But too late to help with our CO_2 emissions."

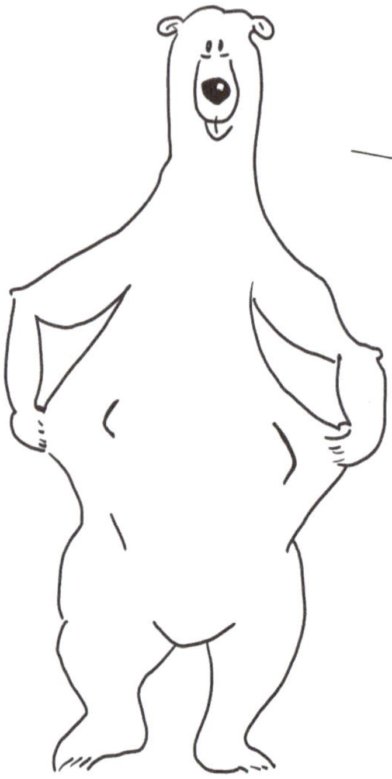

And me, what am I going to do with my sweet, warm and beautiful fur ?
OH NO!!! NO!!!
Tell me, Eifel, you whom I trust, when will we switch over? And to what temperatures ?

When ?

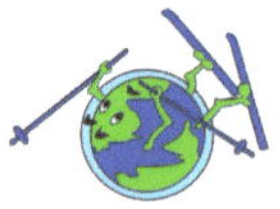

A publication by the University of Copenhagen in July 2023 in Nature (Ditlevsen et al. 2023), which I mentioned earlier, predicts a collapse of the AMOC (roughly the circulation of ocean currents in the Atlantic) as early as 2025, and more likely around 2050, which could lead to an impressive climatic domino crumbling (references 1, 2 and 3 in Ditlevsen et al. Such a tipping could in turn trigger other ones, some of which are probably already underway, particularly in geopolitics..

An additional clue was given in two recent publications (St George, S., 2019, Neukom, 2019) that highlight the peculiarity of the present warming episode. Unlike Little Ice Ages of the previous few centuries, which occurred in turn on different continents, the current warming affects the entire planet, in a synchronous way.

Amazing! That sounds a lot like the critical point in the sheep example you explained earlier

Yes, well done! This is an additional clue that we have indeed reached the immediate vicinity of the critical point! It's actually unbelievably close, in a few years' time. We might already have one foot in it! Will humankind be able to react in time ?

But 2050? 2025? That's tomorrow!!!

Yes, it's tomorrow, as you say!

But what temperature will we reach?

What will be left of my dear ice pack? And what about my seals ?

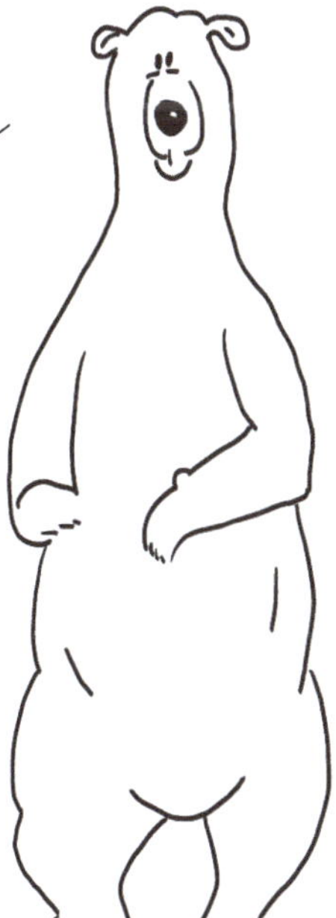

Up to which temperature?

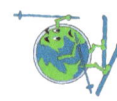

It's hard to say. If we refer to the PETM I already mentioned (56 million years ago) where roughly the same total amount of CO_2 was injected as we are doing now, it ended up to about 10°C above the preceding period, and we may expect comparable consequences.

That's absolutely huge. Few species could survive it! What about bears, penguins, seagulls and humans ???

I mentioned this earlier. Yes, a lot of species, both animals and plants, would have to worry about, just because they are interdependent. And more especially humans. And that would only be fair, because they are ultimately responsible for this huge change in climate!

If I understand correctly, I won't have any more seals to catch in ice pack holes ?

Yeah, it'll be rough !

My cousin Grizzly who lives in Wyoming told me that Yellowstone River salmon are delicious. Do you think they'll make it all the way up here? Unless he invites me to his place ? I'll have to learn how to catch them !!

And to adapt, if you can do it. But going further, it's not just seals or salmon that would disappear.

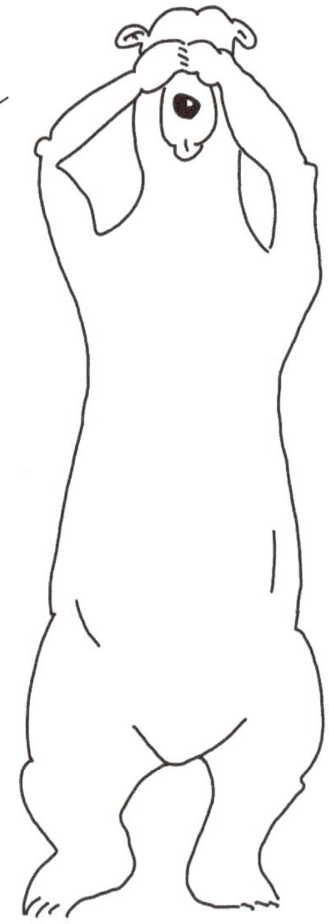

No, no, Eifel, stop, please! I can't believe in it! That's too scary… What if this was all just a "theory" ?

Wouldn't that be nice believing it, Bear?
But remember that Science is not about believing or not believing.
We only share and disseminate what we have carefully observed or rigorously demonstrated, whether it makes you happy or not! The rest is nothing but hypothesis, assumptions or poppycock !

However, I once saw a funny guy who was trying to fly (yes, to fly!), and who didn't seem to believe in a theory that you had already explained to me and that has been verified and demonstrated for a long time, Newton's gravitation, if I'm right.

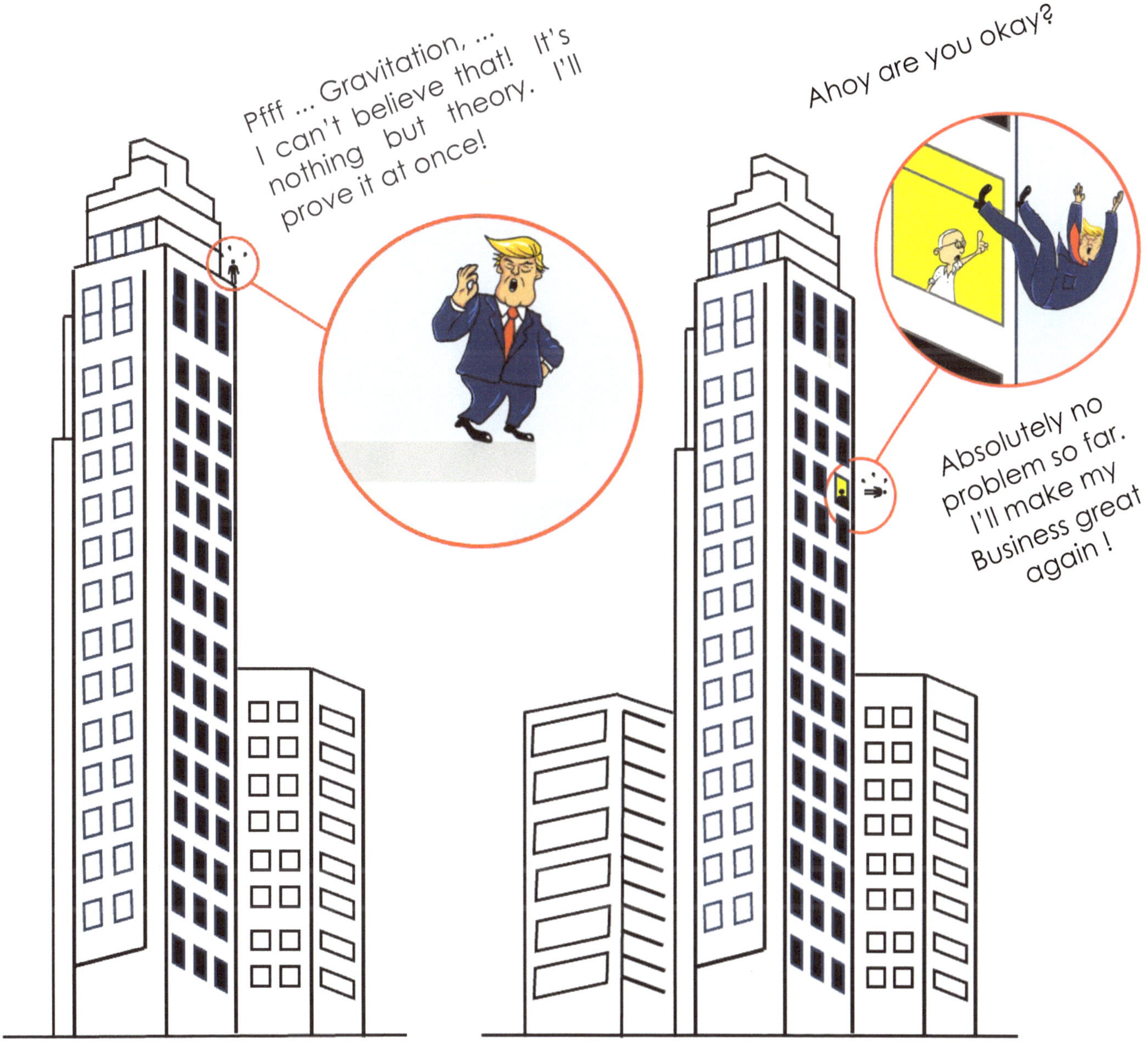

Pfff ... Gravitation, ... I can't believe that! It's nothing but theory. I'll prove it at once!
Ahoy are you okay?
Absolutely no problem so far. I'll make my Business great again !

Yes, some people find it hard to believe what scares them, or what affects their wallet (or their account in the Bahamas).
I hope that Humankind is not there yet, ready to take a further step into the void, and believing there's a way to step back if necessary.

5
And now, what should we do ?

If the only tool you have is a hammer,
any problem will seem like a nail to you.
(Abraham Maslow)

5.1 Where are we staying right now?

A man never rises higher than when he does not know whither his path can still lead him.
(Cromwell)

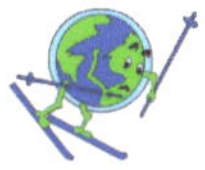

So, you, Eifel, what do you conclude from all this ?

What do I conclude ?

We are in an
ABSOLUTE EMERGENCY

"If you were in a car hurtling towards a precipice
would you say, "I'll put the brakes on
when I get to the edge,
or would you jam your brakes IMMEDIATELY"?
(after a talk by Kate Jeffery,
Professor of Neurosciences.
University College London, 2019)

« Humankind is facing
collective suicide »
2022
Antonio Guterres
Engineer, Physicist,
Former Portugal Prime Minister
UN Secretary General

"AMOC collapse is expected to occur around the middle of this century »
(Peter & Susanne Ditlevsen, Copenhagen University, Nature Comm. Jul 2023) https://veluxfoundations.dk/en/can-we-predict-sudden-climatic-changes

<u>It's imminent</u>. Because the system has inertia, <u>if we don't take drastic action now</u>, it will be too late. When the canoe reaches the edge of the waterfall, you can't go upstream anymore !

By the way, what does AMOC mean ?

I think I've said it before, but I'll clarify. It stands for "Atlantic Meridional Overturning Circulation". It is the complex system of ocean currents flowing in the Atlantic, bringing warm waters from the Gulf of Mexico to European coasts, and then cooling down in arctic zones (that increases water density), diving and finally heading back south in ocean depths. The Gulf Stream, better known, is a branch of it.

Why would AMOC collapse? And how could its collapse influence climate?

That's a fine example of a positive feedback! As climate warms, Greenland ice melting speeds up. This freshwater flows into the ocean, and because it's less dense than saltwater, it hinders AMOC's saltwater sinking into the depths. As a result, AMOC's driving force slows down to a possible complete stop. This, in turn, could accelerate global warming through a series of cascading events, as I'll explain in a moment, enhancing Greenland ice melting rate, and so on...

But very locally, the climate of Western Europe might cool down (or better, warm up more slowly).

Cooler water, more fish! That's good news for fishermen (and for me of course)!

Alas, no. First of all, as I just told you, it won't be cooler, but it may just heat up more slowly...! Secondly, ocean currents form a huge water redistribution system on a global scale, and a possible AMOC collapse, which is an essential piece in climate change (Ditlevsen & Ditlevsen, 2023), would, I repeat, trigger chain reactions all over the globe, and through a gigantic bowling party, would likely lead us to a major climate (and probably general) catastrophe, as I mentioned earlier.

5.2
Between climate control and resilience

A climber is someone who takes his body to the place his eyes have already seen.
(Gaston Rebuffat)

Is CO_2 the main culprit ?
Yes, that's clear! And so, it is absolutely necessary to reduce its production down to zero as soon as possible !!

Provided I'm supplied with all the energy I need, of course!
HAA HAA HAA
And how to do it???
And why do we need to emit so much ?
Basically, the main emitters are heating, transportation, and manufacturing of course. But since electricity plays a key role in these 3 areas, I'll first explain how it is produced. We'll see later on a few examples of how it is used, in combination with other forms of energy.

5.3
How is electricity produced ?

I'm going to make electricity so cheap
that only the rich will have the luxury
of using candles.
(Thomas Edison)

Yes, I really need to tell you how electricity is produced. Currently, it is mainly produced by "thermal" power plants, but also by renewable sources. Let's start with "thermal" ones.

What does "thermal" mean ??

This means that we start by heating water. This water produces steam at high pressure and high temperatures, which is then used for turning turbines, like ultra-fast mills. These turbines then drive alternators, that generate this famous electricity, which is then transported by high-voltage lines.

So far, so good. Sounds awesome !

Awesome ? Not quite.
Sadi Carnot (then 28 years old) had demonstrated in 1824 (Carnot, S. 1824) that the efficiency of these "thermal machines", that transform heat into mechanical (or electrical in our case) energy, was all the higher the hotter the steam was at the motor inlet (in this case, the turbine) and the colder it was at the outlet.
This is Carnot's famous law. That's why thermal power plants need to be cooled down, just like combustion engines in cars.

Sadi Carnot (1796 - 1832)
French Physicist and Engineer

And is that a big deal ?

Judge for yourself!

Under the operating conditions of a thermal power plant, the efficiency is in the range of 30 to 35%. This means that when we produce 1 kWh of electricity, we pour twice as much heat into rivers or seas, which does not help climate control !

Yes, all this waste is not that smart! And so, unless I'm mistaken, heating with an electric radiator is stupidly converting back into heat this small fraction of electricity that this waste has allowed us to produce !

Exactly! You've got it all figured out! Another example: When you hit the brakes to stop a car in motion, they heat up. But if you burn gas under the car to warm up the brakes, the car won't move. That's why, to move the car forward, we usually set up a combustion engine under the hood, that obviously obeys Carnot's law. It therefore also suffers from the same waste, producing mechanical energy while pouring more or less twice as much heat into the atmosphere through the radiator and the exhaust pipe. That's the difference between "disordered energy" (heat) and "ordered" energy (mechanical, electrical,...). In hot matter, molecules move at random in all directions, whereas in a (cold!) moving car for instance, all of them propagate in the same direction.

Fine, I got it! This means that thermal energy and electrical (or mechanical) energy are not equivalent, not interchangeable, and that one of them (electrical or mechanical) is more valuable than the other one (thermal) ?

Yes, that's right.
But there is another drawback to some of these thermal power plants, which is better known. It's their huge CO_2 emissions, at least for those using fossil fuels (coal, oil, gas) that contain lots of carbon.

Why not bury all this carbon underground. Wouldn't it be a good solution to send it back to where it was taken from ?

Be careful, Myrtle, there is often a lot of confusion between Carbon C and carbon dioxide CO_2.
The coal that is extracted from mines is carbon C with varying degrees of purity. Heat is obtained by burning it, i.e. by combining it with oxygen from the atmosphere. But such a reaction produces CO_2, which is the main culprit for greenhouse effect and global warming, and not carbon !

So, we won't solve the problem by burying the coal where we found it, since we've already burned it. And in any case, carbon has nothing to do with greenhouse effect as long as it's not combined with Oxygen to produce CO_2.

OK, yes, and as such, the expressions "carbon footprint", or even worse, "burying carbon underground", are really confusing!

Carbon sequestered underground" is usually called a coal mine, isn't it?
So it's CO_2 that should be buried, instead of C.
But CO_2 is a gas! What to do ?

This is being studied, particularly in Iceland, where scientists are considering injecting it at depth with water into basaltic rocks, with which it would recombine into carbonic acid and then carbonates. But the instability of Iceland's subsoil raises questions.

In any case, it would only concern a tiny fraction of emissions. And this could be a good pretext for continuing to extract and burn fossil fuels. How long would we be able to accumulate and store our waste, CO_2, nuclear, chemical...?

Eifel, you were talking earlier about thermal power plants that emit CO_2. Isn't there anything else besides these terribly polluting thermal power plants? What about nuclear energy?

Well, nuclear power plants may possibly be considered, or at least discussed. But hey, Myrtle seems to have something to say about it !

Just to repeat what I'm hearing everywhere; nuclear power would be the right solution... A lot of energy produced, almost no CO_2, so very clean, it's good for climate, isn't it? Everyone in my herd agrees with that !

Watch out Myrtle! Just because all the sheep in your flock say the same thing doesn't mean it's true.

Sorry, Eifel, I was just repeating...

No problem, Myrtle. But II'd like to insist.
A few centuries ago, many humans had forgotten how the Greek astronomer Eratosthenes measured Earth's radius, and instead were convinced that the Earth was flat and at the center of the world. Those who dared to say otherwise were persecuted and often burned alive!
But we've figured out a lot since then !

Er... it seems there are still people believing in Flat Earth

Yes, the "flat Earth" is a belief that has already travelled more than 3 times around the world!
Almost as many times as I did myself!
When I think that you humans imagine yourselves to be the smartest among all animals... It's enough to make me despair!

Er...may we come back to our topic?

OK, don't worry, Barky. I hadn't forgotten your dear sheep! What you're right about, Myrtle, is that nuclear power plants don't emit much CO_2, which is great. But on the other hand, what is not very well known is that they are ALSO THERMAL power plants.

What? You say: nuclear power plants are thermal power plants ???

Absolutely. Like other thermal power plants, they heat up water to turn it into steam... you know the rest, Carnot, waste, etc. The only difference is the water heating process, based on fission of uranium nuclei rather than on combustion of coal, oil etc.
As Bernard Laponche** has said, this is the most expensive way we have found to boil water. And as a thermal system, I repeat, they also obey the inescapable Carnot's law !

** Graduated at Ecole Polytechnique, Paris, PhD in Nuclear Physics, former engineer at the "Commissariat a l'Energie Atomique" (CEA).

You're kidding? You really mean that a nuclear power plant wastes twice as much energy as it produces as electricity?

Yes, alas, that's right. There is nothing magical about nuclear power!
Carnot's law is a physical law, impossible to repeal! The latest generation of European Pressurized Reactors (EPRs) have an efficiency of around 35 % **. Just as for other thermal power plants. When a nuclear power plant produces 1 kWh of electricity, it releases a waste of 2 kWh which will heat the planet.

**(https://fr.wikipedia.org/wiki/R%C3%A9acteur_pressuris%C3%A9_europ%C3%A9en).

But I thought nuclear plants were carbon-free and did not pollute the planet with greenhouse gases!

Not carbon-free... Just low-carbon. Let's not forget that their "carbon footprint" is linked to uranium ore extraction, uranium enrichment, and storage of an increasing amount of radioactive waste that must be permanently cooled down. Cooling requires additional power consumption, that also contributes to general planet warming.

That's about energy waste. But in addition, as far as pollution is concerned, nuclear power plants produce radioactive waste which can have very long lifespans. Apart from burying, we don't really know what to do with it!
To date indeed, none of the breeder reactors, which are meant to burn at least a small part of radioactive waste while producing electricity, have been working continuously so far.

Sheesh. How else can we produce electric current?

With so-called renewable sources, like waterfalls (hydroelectric power plants) or sea currents (tidal turbines), wind (wind turbines), and solar radiation (photovoltaic panels). etc. In short, everything that directly produces what Penguin described earlier as "precious" energies, not subject to Carnot's law.

But I often hear about biomass, biogas, and ... who knows what else! I'm hoping that no one comes along one day to take my lovely grass and turn it into ... electricity !

You're right, Myrtle. We often hear about biomass, and wood in particular. Burning wood simply for heating is useful as long as you don't burn it faster than it grows. I think you will be able, Myrtle, to tell us about it from your mountain pastures in a moment.

But producing not heat, but electricity by powering old coal power stations (subject to the inevitable Carnot's law waste) with wood (which is not exempt from Carnot's law either) is pure heresy. And to do it with wood from forests ravaged by clear-cutting in Europe, or even worse imported from South America by cargo ships is ecological nonsense. And when you realize that they are often "replanted" with conifers, which can hardly be used to produce energy, it becomes crazy.

I understand. But should you tell us a bit more about those « non thermal » energy sources ?

Yes, let's get back to them. How much is, for renewable sources, this famous efficiency you were talking about, Eifel, that was of about 30 or 35% for thermal plants ?

Highly variable, depending on the energy source, but it doesn't matter much for climate. A hydroelectric power plant, for example, returns water into the river, and at the same temperature, without emitting CO_2, without making radioactive waste, and without heating the atmosphere. The only thing it takes from water is some kinetic energy (due to water motion), that would otherwise have contributed to mountain erosion.

That's an example. Are there other ones?

Yes of course. Let's take the case of a photovoltaic system. The efficiency is currently around 20%, but nothing is wasted. We only take 20% from sun's radiation power (which would have arrived on Earth and warmed it anyway) and transform it into electricity. That's as much less that would have heated the roof and the atmosphere.

It's as if we were worrying about the efficiency of trees as they are producing wood through photosynthesis, consuming CO_2 and solar energy.
They just take what they need. No waste!
That's quite virtuous.

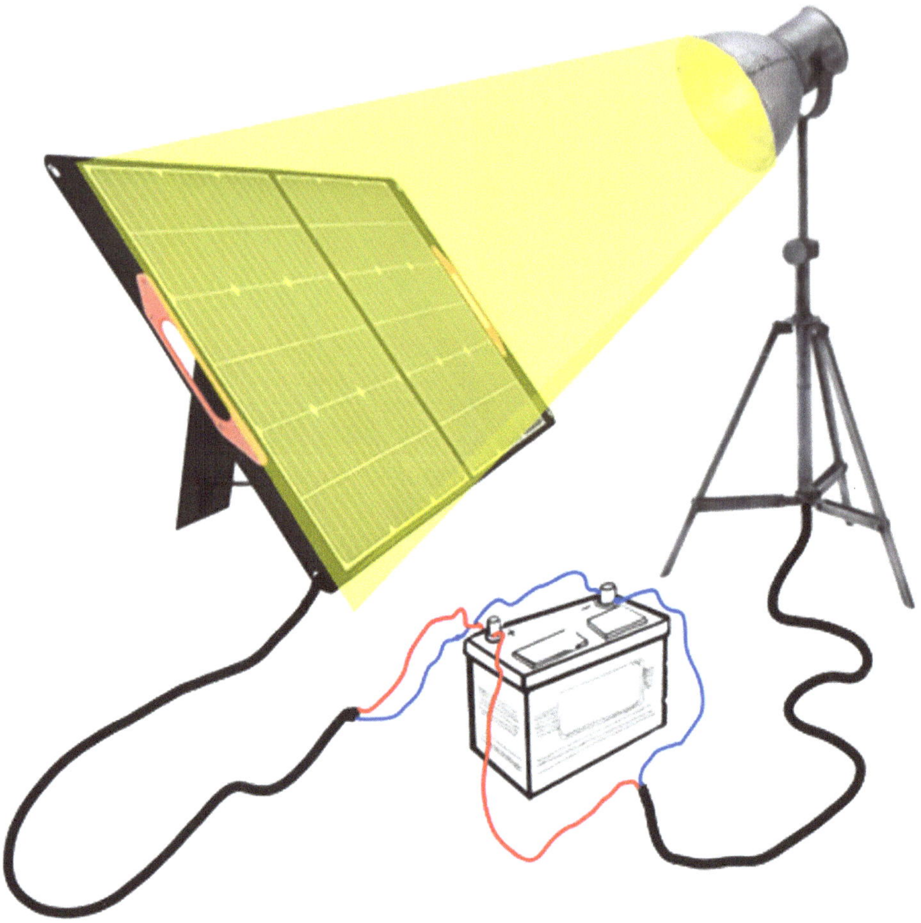

On the other hand, this closed-circuit setup is stupid. Since the panel, lamp and even battery efficiencies are all less than 100%, this kind of installation cannot operate continuously, and will shut down quickly, while finally transforming all this energy into heat! It will be even faster if we try to draw energy to power any device.

One could imagine other absurd ideas, such as for example a wind turbine producing electricity, which would turn a fan producing wind, which in turn would turn the wind turbine, etc...! It's called perpetual motion!

But I've heard lots of criticism about renewable energy, whether solar or wind !

What, for instance ?

They are intermittent and not drivable.

Rain and sunlight, which make our plants grow, are also intermittent and cannot be produced on demand. Does that mean we should give up farming?
On a more serious note, thanks to evolution, the living world on Earth has gradually adapted to these imposed intermittencies. And we now have some technical means to manage them. I'll come back to that in a while.

In summer 2013, while flying above the Rhône Valley in France , I noticed that the activity of the Tricastin nuclear power plants, supposedly non-intermittent, had seriously dropped down. I was told this was because there was no longer enough water to cool them. They had even raised the allowable temperature limit of discharged water to avoid reducing too much electric production.

Trying to replace a physical law with an administrative one never gets you very far!

Soon, all the fish I'll catch there will be pre-cooked!
Barf! Who's ready to taste Tricastin fish chowder?

Yuck! What about my preferred raw, fresh seal meat!

No surprise! To keep these plants producing continuously, getting rid of the 2 kWh of heat for each kWh of electricity produced is mandatory. The cooling system must be permanently available at its maximum efficiency. And the more we produce, the more we heat. Anything else?

Solar panels: "large amounts of agricultural land would be rezoned (France plans to replace agricultural fields and forests with solar panels in areas summing to 3 times the size of Paris)"
(Jancovici & Blain, 2021 p. 160)

Except that... Paris may be the "center of the world" (sorry, of France), it is only 100 km^2. Three times its surface is therefore 300 km^2, i.e. 0.05% (I did say 0.05%) of the country's continental surface area (543000 km^2). One may compare these 0.05% to the 10% of agricultural land in France which has already been rezoned for non-agricultural use. I would add that these 300 km^2 are equivalent to the additional area rezoned in France every year !!

"*Supplying all of France's energy needs would require a wind turbine every km^2! [...] There wouldn't be a single wild place anymore.*"
(Jancovici & Blain, 2021, p. 127)

*Well, just somewhat... exaggerated. That might be true if we wanted to produce all the country's energy (whether electric or not, all combined) from wind power alone, which would be technically, economically and physically absurd. On the other hand, if we consider producing all the electrical energy consumed (475 TWh/year ** in 2019, i.e. an average power of 54 GW**) using only wind power (which will never be the case anyway), the result would be ... somewhat different!*

** 1 TW = 1000 GW, 1 GW = 1000 MW, 1 MW = 1000 kW, and the same for TWh, GWh, MWh and kWh

First, considering basic standard wind turbines, typically 2MW, with a load factor of 30% (proportion of the time when they are active), spread over the 543,000 km^2 of metropolitan France, it would result on average in a wind turbine not every km^2
but every 6 km^2!

But, going further, the new off-shore wind turbines, from the Danish manufacturer Vestas for example, thanks to their height of almost 300m, benefit from more powerful and stable offshore winds, giving them load factors of more than 60% (more than 2 times more), associated with a nominal power of 15 MW (almost 8 times more). So, we save almost a factor of 16 on the number of wind turbines! And offshore, they are not a nuisance, neither for the inhabitants who live at their foot (if any!!!), nor even for the marine fauna which, to everyone's surprise, like to nest in the submerged parts of their structures.
And since wind turbines will only produce part of this energy, the above estimate of wind turbine density will have to be drastically revised downwards!

So, we shouldn't worry too much about intermittency of wind energy or the amount of space taken up by windmills, if I understood your arguments.

Of course not, especially since demand is also intermittent, between day and night for example. A possible problem would be that the intermittency of supply and demand do not necessarily coincide.

But with widely used interconnections between gas, oil, nuclear power plants, etc., we know how to do it, and momentary underproduction in one place is balanced by overproduction elsewhere! And we know how to do that for renewables too!

Yes, without any doubt! But my shepherd doesn't care about intermittency. He prepares his fire only when it's cold. And his wood, which he constantly renews, is not intermittent at all, and it is quite controllable, as you say in human terms !

As they say in my family, the solution is not to put all your eggs in a single basket, and to share baskets with other people..

There will always be places with wind, others with sun, others with both, and everywhere and all the time, available as long as you know how to manage it. And there's biomass, including wood, as you said Myrtle, which obviously needs to be well managed. We also need to play on interconnection,
as I mentioned just now.

A new submarine cable will connect the UK and Denmark to offshore wind farms in the heart of the North Sea. It is in this spirit that true energy sovereignty will be achieved in Europe.

So, we're spoilt for choice. What should you bet on, Eifel? Anything carbon-free I guess? But I hear some people talking about nuclear power again, at least to ensure a smooth transition !

Isn't nuclear power still better than coal with regard to the greenhouse effect?"

Yes, of course, and it's necessary for now.
Indeed, **THE ABSOLUTE EMERGENCY IS TO STOP CO_2 RELEASE**. But it wouldn't happen overnight, and it's a matter of timing.

Is it a question of timing? What do you mean, Eifel ?

Regardless of what one may think of nuclear energy, reactors whose construction begins today will not be operational for another 15 or 20 years...., or even more for the supposed 14 EPRs planned in France for instance. That means they will not even start contributing to reduction of CO_2 emissions until 2040 or 2050. As I said above, there is a very serious "chance" that the climate will have tipped before. And so, such investments would serve absolutely no purpose, except to waste money that could have accelerated renewable energies set up !

So, no nuclear power in your opinion ?

That's not what I said, Penguin. Let's not be dogmatic! Nuclear power is necessary, but on the short term only. The long-term solution is obviously renewable energy, as I explained earlier. They are significantly faster to install than nuclear power plants. But **WE MUST STOP CO_2 EMISSIONS AT ONCE** to avoid irreversible tipping. Except if we live in Scotland, in Denmark, or in some other wise countries in the world, at the current rate, it will take us several years before production would mainly rely on renewables.

And in the meantime, what should we do ??

We have no choice, we no longer have a choice: we MUST STOP BUILDING new nuclear power plants, but we should extend the life of existing plants as much as possible, and then gradually close them at the rate of increase in renewables. And the key word remains: ENERGY SOBRIETY again and again, concentrating as much new investment as possible on renewables.

Sobriety, sobriety, don't you think you're going to scare people by saying things like that?

No, not really! Sobriety doesn't necessarily mean deprivation! Are we really happier driving around in a heavy, polluting 4WD car rather than in an average sized car, as Wolfgang Cramer puts it?

Pfff…Our retirement age has been again delayed!

But still! ... A little birdie told me that that another kind of nuclear power plant is possible: fusion reactors, which would produce a huge amount of energy.

Yes, you're right. These reactors would run as very small suns on Earth, in which hydrogen nuclei (protons) are forced to merge together, producing helium nuclei and a huge amount of energy in the form of heat and radiation. This is called "nuclear fusion". This is the same principle as the H-bomb but kept under control. Unlike the bomb, there is no risk of explosion, but the difficulty is to keep it running. It is necessary to reproduce at very small scale in the reactor the conditions of extreme temperature and pressure (plasma) naturally maintained in the sun thanks to its enormous gravitation (seen from our modest point of view). This is not the case at the moment, at least for reasonably long operating periods...

It should be furiously hot in there !!!

Sure! More than 70 million degrees. I did say MILLIONS! You've got a point there! Such temperatures are associated with incredibly powerful radiations. Imagine the materials making up the reactor vessel. You might wonder which material, no matter how efficient for other purposes, could withstand radiation damage due to the very high energy radiation emitted by a fusion reaction in continuous operation for long. That material would experience a disorganization of its crystalline structure, resulting in swelling and embrittlement (e.g. Victoria et al., 2001, Knaster et al., 2016). The sun sends us similar radiations of course, but we are protected against them by its distance (150 million km), and by high atmosphere and Earth's magnetic field. If some day we manage to keep the reaction active for "industrially compatible" times, I think we'd have to worry about the vessel sustainability !

By the way, I'd like to tell you an interesting anecdote about this problem. One of my best friends was a researcher, theoretician, at Princeton, in the United States. He used to meet specialists there who were working on the Matterhorn project of controlled nuclear fusion. One day he asked one of them, whom he knew quite well, when in his opinion-controlled fusion would be operational. "In 20 years," his colleague replied. And every year, my friend used to ask him the same question. It had become a ritual between them. And the answer was the same : "I've already told you, in 20 years" !

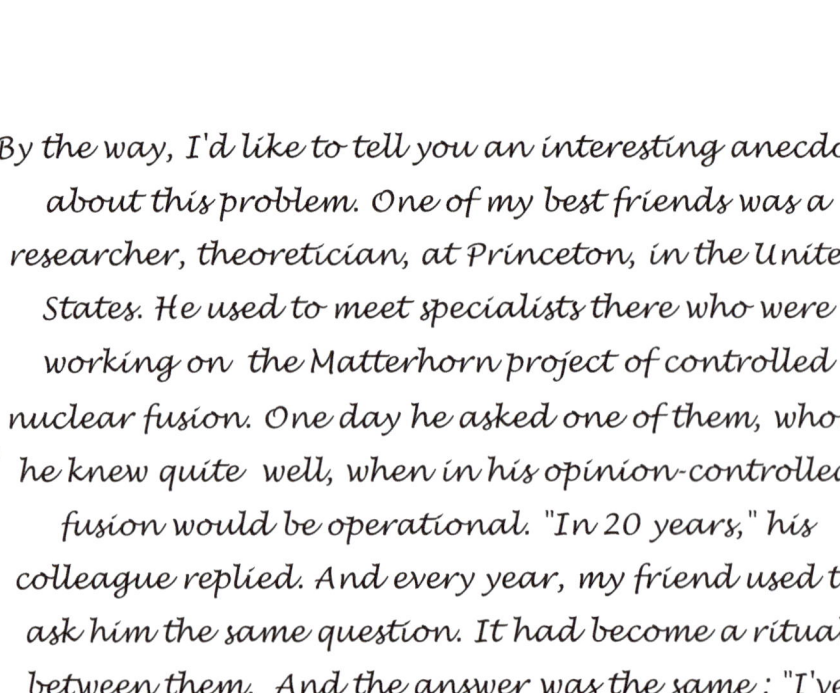

I'd also like to mention that nuclear fusion-based power plants would also be "thermal" power plants, subject to Carnot's law, and to the corresponding energy waste, to which would be added the energy needed for plasma containment, as well as for vessel cooling .
If such plants ever succeed in producing considerable amounts of energy as expected, they will considerably warm the planet anyway !!

Finally, a final remark, more general. In a chaotic and disrupted world, exposed to all kinds of climatic, industrial, agricultural, economic, geopolitical accidents etc ... it is essential to optimize resilience. Or, as biophysicist Olivier Hamant recently said (Hamant 2024), resilience and efficiency are incompatible. A system that is too 'pointed' will be extremely fragile. « Robustness is playing in the gears » Accepting to lose efficiency to gain robustness will make the system much more adaptable.

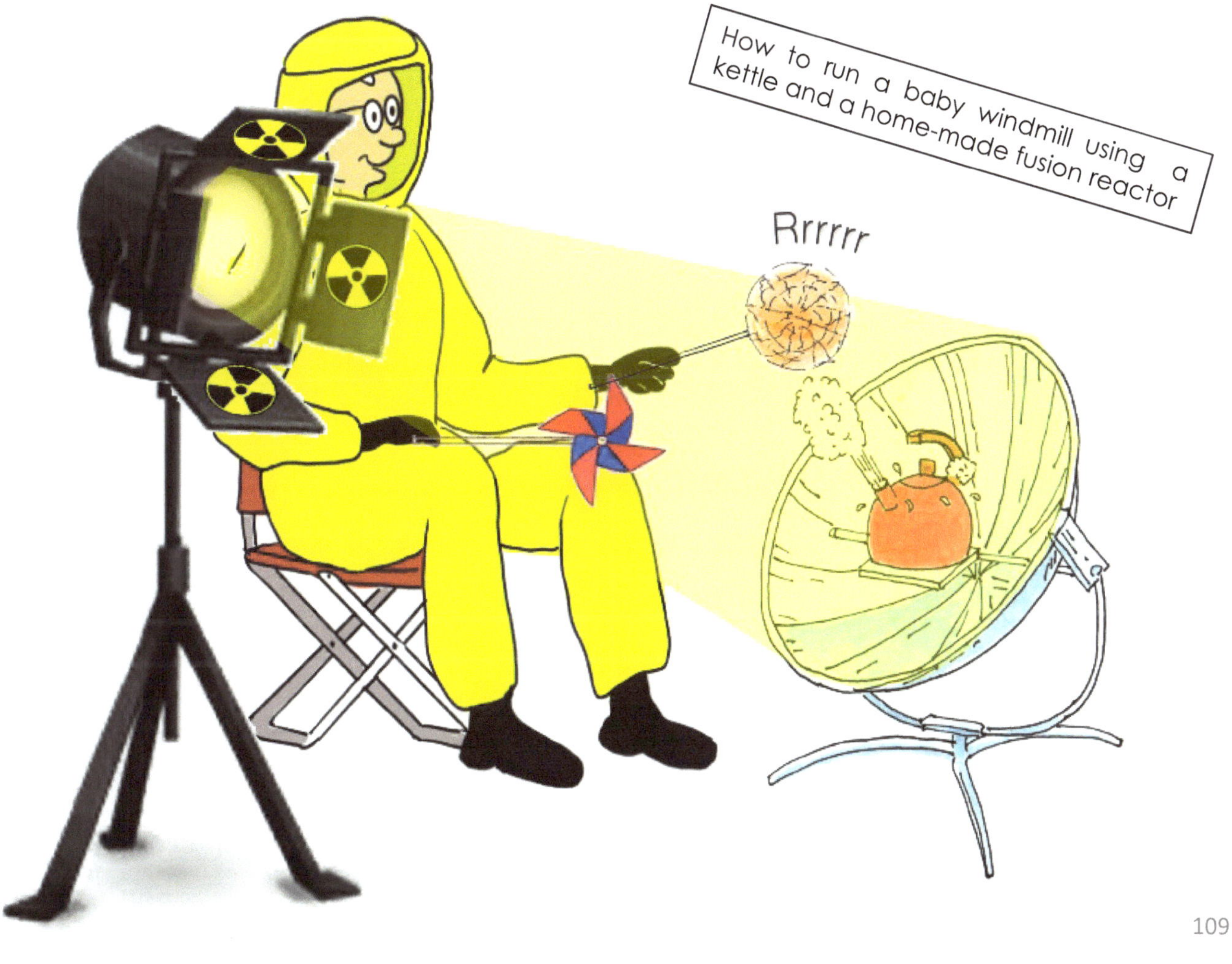
How to run a baby windmill using a kettle and a home-made fusion reactor
Rrrrr

5.4
Heating

Cold is the pain of believing
that heat will never come.
(John Berger)

Raining today
Is better than on a sunny day
(Pierre Dac)

Heating ??? Just a moment, Eifel! You've been telling us for a while that the planet is warming too fast, Bear and Penguin told us how they are experiencing this drama at home, and you're going to tell us about ... heating !

You're right, Seagull! But what I was talking about was the increase in the average temperature on the planet. You who travel a lot, you know that when it's summer at Bear's place, it's winter at Penguin's. The same goes for people who live in Europe or North America, and those who are in Australia, Southern Africa or Argentina. If it is 55°C instead of 45°C in Rosario or Mendoza in January, it will perhaps be 0°C instead of -10°C in Berlin, and Berliners would still appreciate heating...

I can understand. But even if the bear is the symbol of Berlin, that doesn't mean I'll sell them my fur! !

If they don't have your fur, they'll have to warm up. So, how? When it's cold, we huddle together, and we've got wool on our backs! But our shepherd, for his part, comes back to his hut and lights a wood fire in his stove.

He's right. He has lots of wood next to him, and he only cuts what he needs, without contributing to greenhouse effect.

... but...! I thought that wood, when burned, produced CO_2!!

Right again, Myrtle. But the CO_2 that the wood releases when it burns is the CO_2 the tree has absorbed during its growth. The balance is neutral.

OK. But the coal that we criticize so much is also carbon stored by plants that grew some time ago, during the Carboniferous period, roughly 300 million years back. Why shouldn't this balance still be neutral ???

300 million years?
My God!
Flowers did not yet exist
at those times.
Poor Meganeura, forced to feed
on disgusting cockroaches and
grasshoppers!

I admire your erudition, dear Seagull!
Actually, it's again a question of time scale. The wood that the shepherd is burning today is wood that grew during the last 20 or 30 years, and the CO_2 he produces by burning it feeds trees that are growing at the moment, collecting the carbon to make wood and evacuating the oxygen we breathe into the atmosphere thanks to sunlight (photosynthesis).
That is wood that he or his successor will burn in 20 or 30 years. It's neutral as long as it doesn't burn faster than it grows !
On the other hand, the large amounts of carbon stored during the Carboniferous period (that lasted roughly 60 million years) and have been burned massively since the beginning of the industrial era (only 150 years!) will not be renewed anytime soon on a geological scale, and even worse on a human scale, and this balance is not neutral.

And electric heating maybe? I've heard that it doesn't emit CO_2.

Again, Myrtle, please don't believe everything you hear without thinking it through. And you may not have listened Penguin just now, who understood that electricity was a precious energy, or Seagull, who explained that electric heating was an aberration, a huge energy waste!
But there's a way out. This is the heat pump. It does not produce thermal energy, but simply extracts it from outside and pours it inside. It's the same principle as a refrigerator, although working in opposite direction.

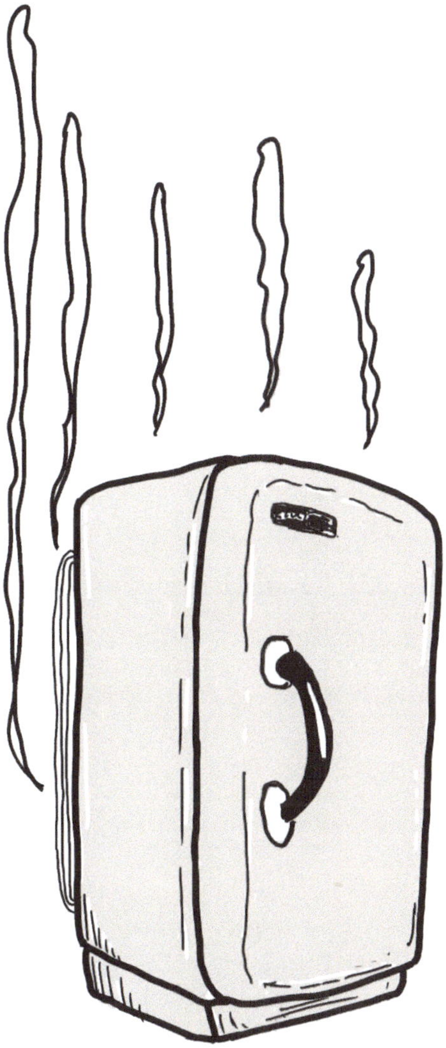

You mean we'd go get heat
where it's cold and put it where
it's not so cold??? Is it like
cooling the fridge to heat up
the kitchen ?

It's not "as if." That's exactly what's happening in the kitchen. A heat pump is like a pump that brings water up from the well. We don't create heat, but we pump it from outside (where there is always some left) to inject it inside. This makes it possible to short-circuit the waste of thermal power plants of all kinds, due to Carnot's law. It still consumes a small amount of electricity, but far less than an electric convector.

Pfff, heating and heating again, that's all we're talking about! For my part, I'd just prefer the opposite, especially in summer !

Good thinking, Bear! It turns out that most heat pumps can run backwards in summer, which electric heaters can't! It's called air conditioning. It's often criticized, as it consumes some electricity, and blows hot air towards the next buildings across the street when you live in a city. But if you are staying in less dense areas or in the "countryside", and if you are equipped with photovoltaic panels for instance, you'll be often in excess of solar power in summer, which is a good thing, since it gives you a free "air conditioning" without bothering anyone.

But but... There is one intriguing thing, though. We are constantly told about poor buildings thermal insulation, that we need to improve and improve again.. Apart from experimental aerodynamics, I don't know much about physics, and my question will probably sound stupid to you. Does insulation that protects against cold in winter also protects against heat in summer ?

No, your question is not that stupid, and I should say quite interesting! In principle, yes, it's theoretically symmetrical, but in practice it's not that simple. It would be necessary to play with both insulation and thermal inertia.
In winter, even if we keep a temperature at 19 °C thanks to our heating systems, it will always be colder outside than inside, day and night. Preventing heat escaping through good insulation always means saving as much heat.

In summer, it would theoretically be the same thing in the opposite direction. If you turn your heat pump to the cooling function (air conditioning), insulation will prevent outside heat from entering your cool home as it prevented it from escaping in winter.
But in practice, it's not symmetrical. First, the widespread use of air conditioners in cities would make heat unbearable outside buildings.
But above all, though in "still temperate" latitudes most homes are equipped with heating, the same cannot be said for air conditioning, if only because cheap summer solar electricity to feed them is still not widely spread.

Without air conditioning in summer, if it's hotter outside than inside during day, protection against outside heat is useful. But external temperature may be somewhat lower at night, at least in sparsely urbanized areas. High insulation, too few and small windows and too large thermal inertia may prevent from cooling the interior by taking advantage of cooler external night temperatures.

Sure, I couldn't imagine being locked up between four walls, without any opportunity to dive into cool water, or feel the wind flowing on my feathers !

Not everyone is so lucky, Seagull! In any case, in the current climate perspective, it is essential not to reproduce or extrapolate old patterns. We need to imagine new, modular and smart solutions, playing between insulation and thermal inertia, combining winter insulation through innovative techniques and summer night ventilation that could be inspired by the traditional architecture of tropical countries.

If you don't live in the middle of the forest or at the foot of a waterfall, you can opt for clear reflective roofs, white or vegetalized facades, wide roof overhangs, flysheets (that may be equipped with photovoltaic panels) allowing circulation of cooler air during summer nights, (or production of sun preheated air for heating during beautiful winter days). And why not structures on stilts that could be closed in winter to store firewood ?

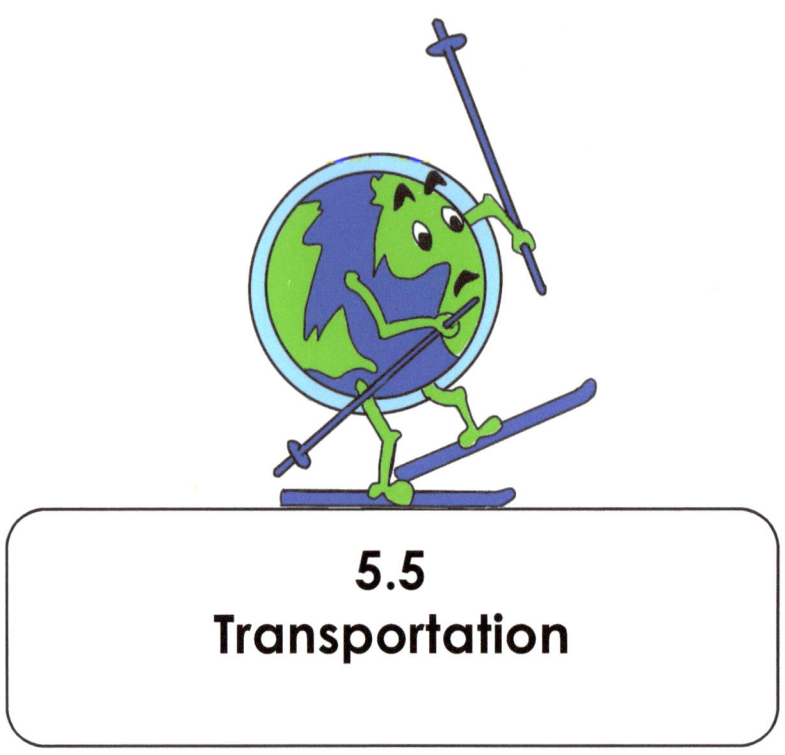

5.5
Transportation

"Innovative responses are local, because local allows for disobedience. Innovation is successful disobedience."
(Jean-François Caron, Mayor of Loos en Gohelle, France)

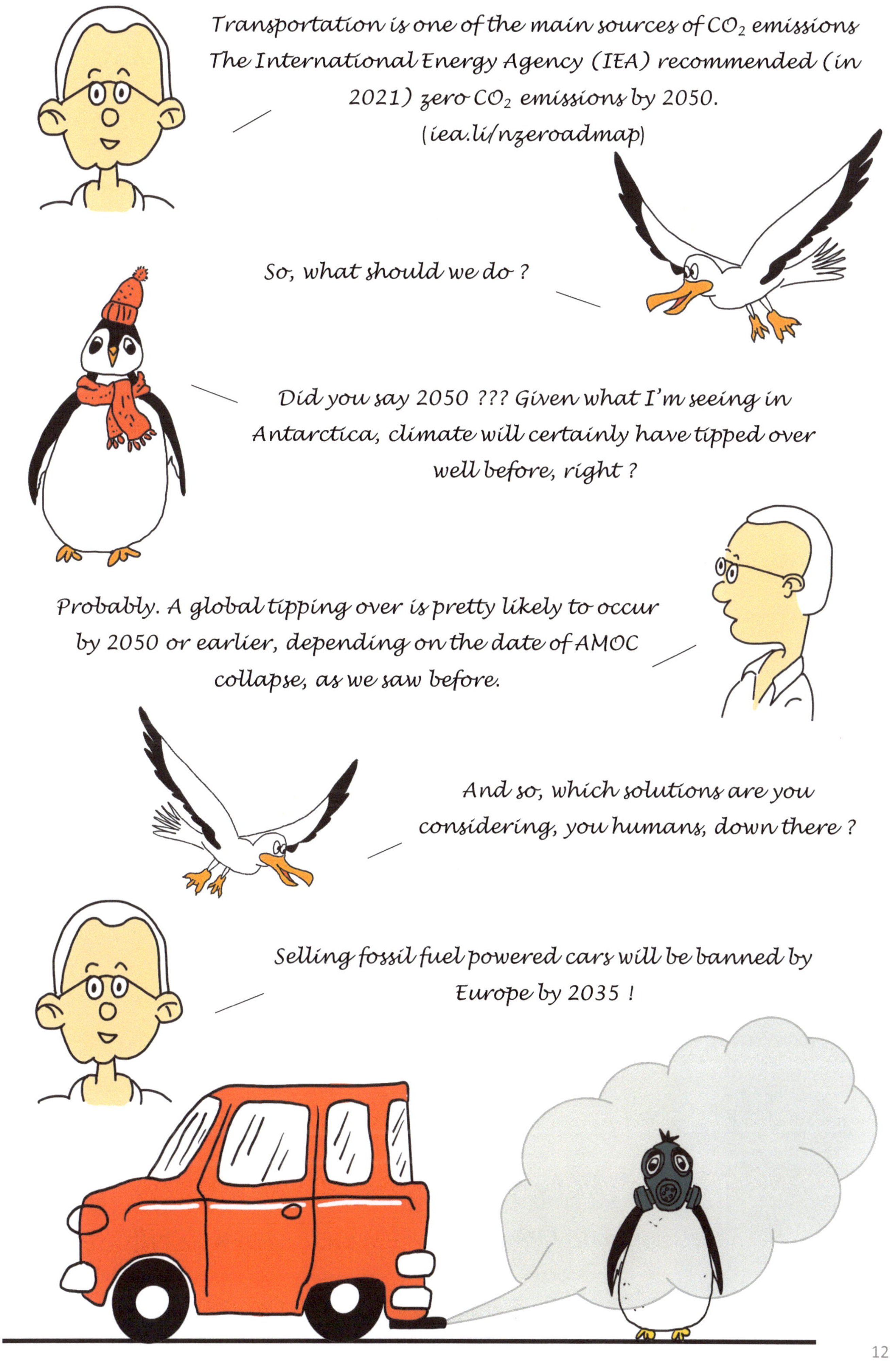
Transportation is one of the main sources of CO_2 emissions The International Energy Agency (IEA) recommended (in 2021) zero CO_2 emissions by 2050. (iea.li/nzeroadmap)
So, what should we do ?
Did you say 2050 ??? Given what I'm seeing in Antarctica, climate will certainly have tipped over well before, right ?
Probably. A global tipping over is pretty likely to occur by 2050 or earlier, depending on the date of AMOC collapse, as we saw before.
And so, which solutions are you considering, you humans, down there ?
Selling fossil fuel powered cars will be banned by Europe by 2035 !

I don't mind cars very much. But planes ?
I always see so many of them passing over
my mountain pastures.
Don't they pollute?

It depends on what you're comparing to. An airliner burns roughly 3 Liters of fuel per 100 km per passenger. A standard combustion engine car burns about 6 litres. Therefore, a single person in a car emits twice as much as if he were in a plane!
Hence the interest in carpooling when possible.
But of course, one can hardly fly to work, or drive a car across the Atlantic or the Pacific ocean.

And so, banning long plane journeys is a good thing !

Obviously. In addition, what I was saying relates only to airliners carrying a large number of passengers. The consumption per passenger of "private jets" is scandalously higher !

As long as you humans can't build clean airplanes, it's going to be a big deal.

We'll probably get there one day. But again, it's a time issue. Climate tipping must be avoided RIGHT NOW! Otherwise, what would remain of Humankind will no longer be in a position to invent or build anything.

Pedal plane, probably the cleanest aircraft in the future !

And you, Seagull, who never stop flying, aren't you ashamed of yourself?

Not at all! It takes a lot of energy for humans to stay in the air. Even worse for you sheep, when you jump off a cliff. But I don't have that problem. Millions of years of evolution have given me a light body, perfectly designed articulated wings and "turbulence-free" feathers, along with a thermal engine of course, but with a fairly good efficiency, and running on renewable energy, at least as long as there are some herrings left !

Well, actually, I don't think flying is an option for me. But neither are cars. There aren't any roads around here, much less subways. And swim-pooling is out of the question, since the houses we have on sea ice are scattered so far apart and permanently moving!

And what about electric cars ?

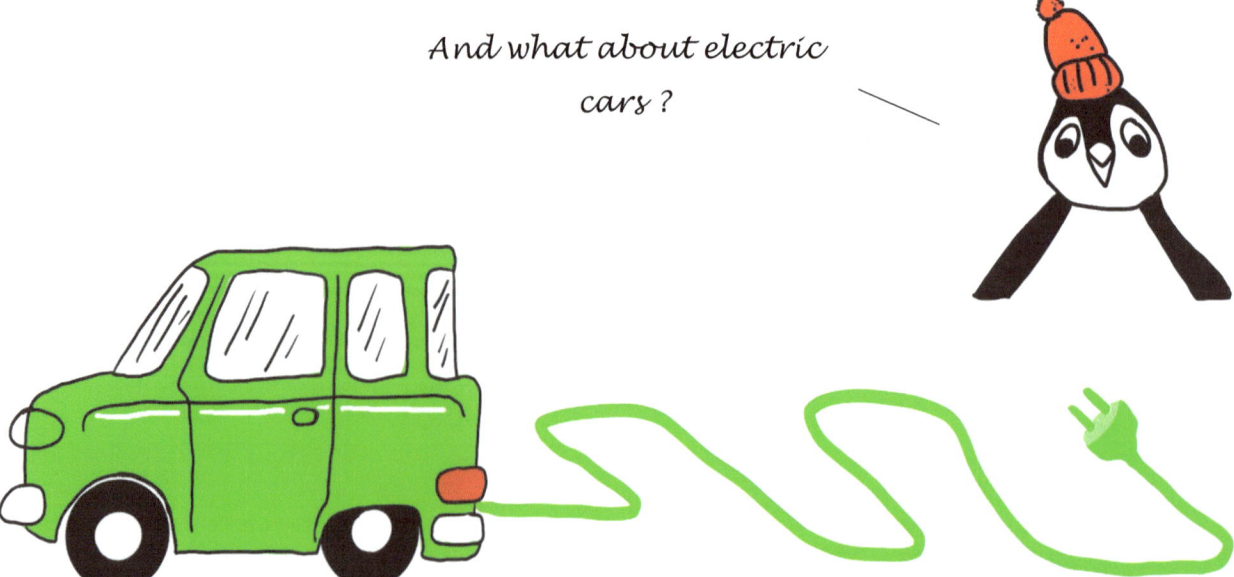

Electric cars ? My Grizzly cousins told me they are all the rage now around their dwellings in Wyoming. There's a guy over there, in California I think, who sells a lot of them.

Yes, I know. But that guy also sells rockets, which aren't exactly electric! Once when I was flying in that area, one of them practically exploded right on my beak!

You're kidding?

For real! It was one hell of a scare, you can 't imagine!
And the cloud of pollution...! Unbelievable!

But electric car operation is not subject to
Carnot's law. So that's clean, right ?

Not necessarily. Think about this way: If the electricity which runs the car was produced by a thermal power plant (including nuclear), then the power plant would have already wasted energy and seriously heated the planet, in keeping with Carnot's law (in addition to pollution). But at least you would not have to waste even more to run the car's engine!

On the other hand, if the car is really using green electricity (generated from hydraulic, wind, or solar energy), then it would not have produced the waste required by Carnot's law. So, it would be running on fairly clean energy. Unfortunately, at the moment we do not have enough green electricity available to make this happen.

And the sail boats I often see where I live ?

It's even more direct, because sailboats don't use electricity at all. But they are dependent on the availability of wind. And you need some basic sailing knowledge.. Right, Bear?

Actually, the biggest obstacle is that electricity doesn't just fall out of the sky. And there's no such thing as a mine of electrons."

* Doc. Emmett Brown (Christopher Lloyd) - Back to the future. A film by Robert Zemeckis - 1985

And wouldn't e-bikes be a great idea?

First of all, regular bikes are a great idea. But sharing space with cars can be a problem. Electrically-assisted bikes consume so little energy, that it would be a shame to miss them out. Not to mention ultralight 2 and 4 seat electrically assisted micro-cars with battery and solar panels on the roof, that already exist!

And also, trains, trams... and cable cars of course. Operating a cable car needs almost no energy, even in rugged terrain, where the energy gained from going downhill zeroes out the energy used for going uphill. In flatter terrain, like cities, it's not even an issue. Plus, cable cars do not need bridges or tunnels, their environmental footprint is negligible, and set up as well as operating or dismantling costs are minimal in comparison to trams.

They are worth taking a closer look at. Right now, there's no such thing as a hydrogen mine which means that we have to make hydrogen with ... electricity! This is called green or grey hydrogen depending on whether it's made with renewable energy or non-renewable energy. In both cases, the process needs a lot of electricity.

Now, could a hydrogen mine actually exist? The hydrogen molecule H_2 is so small that it sneaks in everywhere - that's what we call "diffusion". If, like coal and oil, hydrogen was once stored in deep geological layers, it has evaporated since then. So it is hard to believe that significant amounts of "native hydrogen" exist to the same extent as coal and oil do.

On the other hand, ongoing research in various areas of the planet, and particularly in the Lorraine mining basin in France, has revealed the presence of hydrogen with concentrations starting at almost zero at ground level, but increasing to non-negligible amounts at 1000 m depth (Mining.com, 2023). This is called White Hydrogen. It may result from complex, still-active, high-pressure, and high-temperature hydrothermal reactions between water and siderite (iron carbonate $FeCO_3$)(Milesi et al. 2015). Strictly speaking we might call this not "native" hydrogen, but "nascent" hydrogen".

But we would still have to assess how much of this nascent hydrogen is available as well as the rate at which it can be renewed. Since we are in a hurry to solve earth's warming, the delays required to verify these findings, develop and set up operations for hydrogen mining all seem prohibitive.

That's wild! "White" hydrogen is not available for the time being. Producing green or grey H_2 needs lots of electricity, "clean" electricity still remains scarce, and, despite this, we are driving up demand for electricity with the development of both electric and hydrogen vehicles!"

COP 28 Dubai: Nuclear power generation must triple by 2050 !

And again, 2050 is far too late! New nuclear power plants would not be operational for another 15 to 20 years., as I said before. By the time we reach 2050, the climate change will probably have switched over.

5.6
Biodiversity and food

Some are born male or female,
other mosquitoes, redwoods, or frogs,
and no vanity can be derived from that.
(Bernard Chevassus-au-Louis,
tribute to Hubert Reeves. Oct 16, 2023)

The world is big enough to satisfy the needs of all, but it is too small to satisfy the greed of a few
(Gandhi)

The tiniest aphid which bumps into the web shakes up the whole system.
Strangely reminds me something !
Yes, but this time, it's not my fault that the system has become critical !
No, not your fault, Barky! But it is indeed critical. And so, if I understand correctly, everything is related to everything else! For example, if there are no more seals, there will also be no more polar bears !

In the same way, if there are no more fish,
there will also be no more penguins.

Or seagulls !

And if there are no more penguins,
there will also be no more orcas, etc...!

According to migratory birds I meet while flying over temperate countries, there are also fewer and fewer bees. Fewer bees means less pollination, fewer flowers, less fruit, fewer seeds. They have to go further and further to find something to eat !

Less flowers also means less grass.
Not very comforting for us.
I'll have to talk to my shepherd about it !

My bear cousins from Mongolia don't have enough to eat either. They are forced to change territory, which does not really please shepherds there, whose herds are getting eaten !

No longer than Mongolian cattle itself, by the way!

As Myrtle just said, it reminds me something too!

That's what biodiversity is all about. It built up very slowly over the course of evolution, allowing all species to differentiate in turn, develop, and then vanish, thanks to recurrent mutual adaptations in symbiosis with all the other ones. This evokes the concept of self-organized criticality (Bak et al. 1987) well known *in Theoretical Physics.*

But the sudden disappearance of a single species may jeopardize, through "domino" cascades of unusually large sizes, the overall operation of the whole system, that becomes momentarily "supercritical", with unpredictable and potentially dramatic consequences..

In such a context, the sudden 63% drop in flying insect population observed in the United Kingdom between 2021 and 2025 is particularly worrying, and may be considered as a warning signal for a general tipping point even closer than we might imagine.

LEAPING INTO THE UNKNOWN !

6
Conclusion

If I seem unduly clear to you,
you must have misunderstood what I said.
(Alan Greenspan)

Well, so, I understand it's time to conclude!

If you put a frog in a pan
and slowly increase the fire,
It won't jump, but it'll enjoy the nice warm bath
until it's cooked to death.
We humans seem to do much the same thing

Jeff Goodell
American writer
essentially involved in
environmental issues

I've said it before, but I'll say it
again: humanity is facing a
collective suicide" (2022)

Antonio Guterres
Engineer, Physicist,
Former Prime Minister
of Portugal,
Secretary-General
of the United Nations

"If governments were serious about the
climate crisis, there would be no new
investment in oil, gas or coal right now.
STARTING THIS YEAR
(and that was in 2021 !)

Fatih Birol, Executive Director of
International Energy Agency (IEA)

Maybe you humans should stop procrastinating and start tackling the problem you have created!

Epilogue

A sick planet runs into another one,
who says to it:
"You've got humankind !
But don't worry, I got it once.
It won't last!"
(after Hubert Reeves)

Currently Man is fighting a war against
Nature. If he wins it, he's lost!
(Hubert Reeves)

Only those who are crazy enough
to think they can change the world
achieve this
(Henri Dunant)

Bowling party? Believe me, If you make it You'll be lucky!
Don't be so negative, PenguinS! Remember what Emerson said!

Don't go where the path may lead.
Go where there is no path and leave a trace.
(Ralph Waldo Emerson)

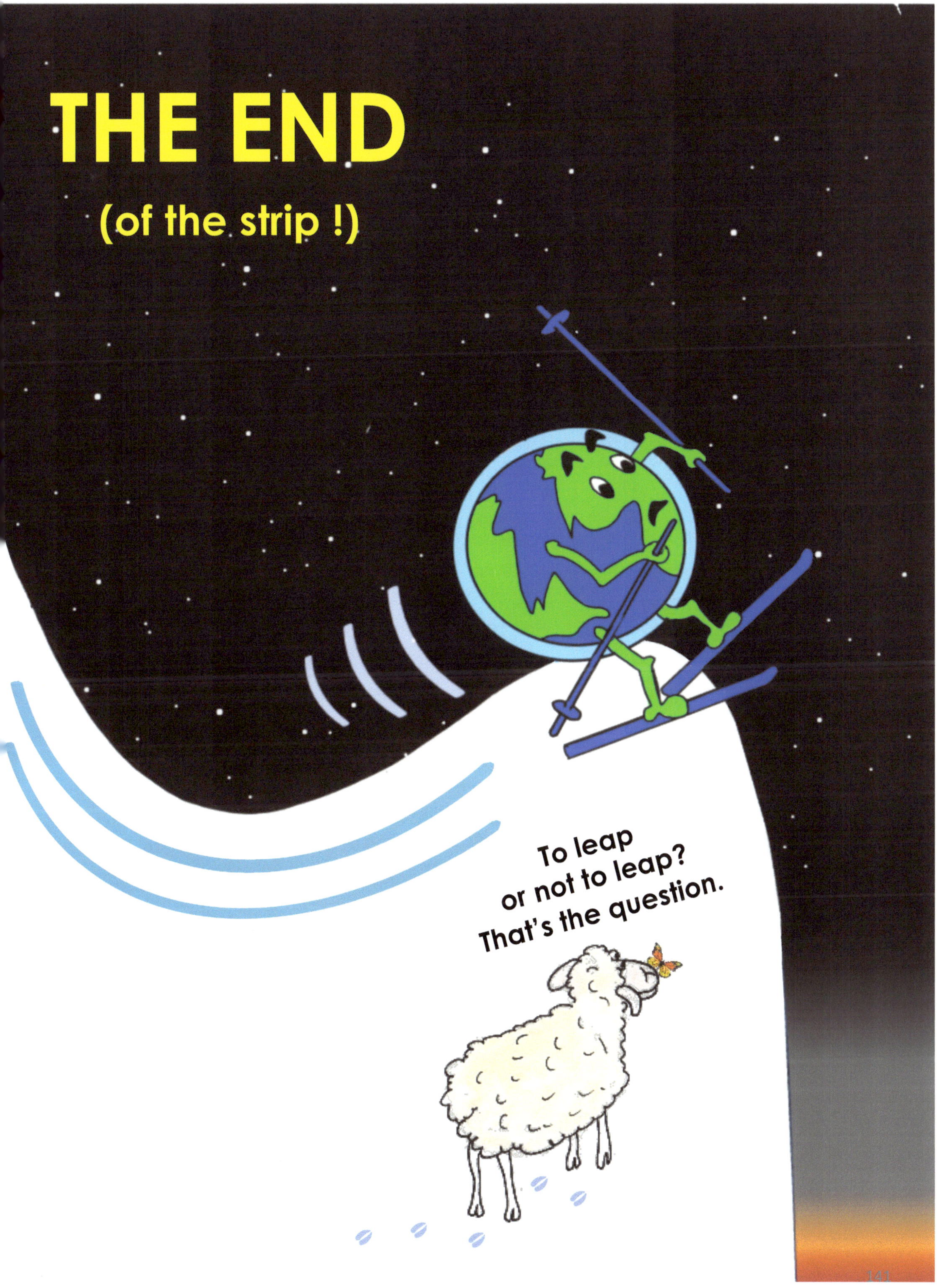
THE END
(of the strip !)
To leap
or not to leap?
That's the question.

References

Alley, K. E., Wild, C. T., Luckman, A., Scambos, T. A., Truffer, M., Pettit, E. C., Muto, A., Wallin, B., Klinger, M., Sutterley, T., Child, S. F., Hulen, C., Lenaerts, J. T. M., Maclennan, M., Keenan, E., and Dunmire, D.: *Two decades of dynamic change and progressive destabilization on the Thwaites Eastern Ice Shelf*, The Cryosphere, 15, 5187–5203,https://doi.org/10.5194/tc-15-5187-2021, 2021.

Bak P., Tang C., and Wiesenfeld K.: *Self-organized criticality: an explanation of the 1/f noise*, Phys. Rev. Letters 59, 381-384 (1987) http://dx.doi.org/10.1103/PhysRevLett.59.381

Bathiany, S., Dijkstra, H., Crucifix, M., Dakos, V., Brovkin, V., Williamson, M. S., Lenton, T. M., and Scheffer, M.: *Beyond bifurcation: using complex models to understand and predict abrupt climate change*, Dynamics and Statistics of the Climate System, 1, https://doi.org/10.1093/climsys/dzw004,dzw004, 2016.

Carnot, S., *Réflexions sur la puissance motrice du feu et sur les machines propres à développer cette puissance*, Bachelier, Paris (1824). Republished in Annales scientifiques de l'E.N.S. 2e série, tome 1, p. 393-457 (1872), http://www.numdam.org/item?id=ASENS_1872_2_1__393_0

Casado, M. et al. *The quandary of detecting the signature of climate change in Antarctica. Nat. Clim. Change* https://doi.org/10.1038/s41558-023-01791-5 (2023).

Ditlevsen, P. and Ditlevsen, S., *Warning of a forthcoming collapse of the Atlantic meridional overturning circulation*, Nature Communications, 25 july 2023, https://doi.org/10.1038/s41467-023-39810-w

Durrell. G. *My family and other animals*, Penguin Books 1959.

Faillettaz, J., Funk, M., and Vincent, C.: *Avalanching glacier instabilities: review on processes and early warning perspectives*, Reviews of Geophysics, pp. 203–224, https://doi.org/10.1002/2014RG000466, 2014RG000466, 2015.

Faillettaz, J., Funk, M., and Vagliasindi, M.: *Time forecast of a break-off event from a hanging glacier*, The Cryosphere, 10,1191–1200, https://doi.org/10.5194/tc-10-1191-2016, 2016.

Holmes, P.: *Poincaré, celestial mechanics, dynamical-systems theory and "chaos"*, Physics Reports, 193, 137–163, 1990.

Jancovici, J-M. & Blain, C.: *Le monde sans fin, miracle énergétique et dérive climatique*, Dargaud, 2021

Knaster, J., Moeslang, A. & Muroga, T., *Materials research for fusion*, Nature Physics, 424-434, 2016.

Lecroart, E., and Ekeland, I., *Urgence Climatique*, Casterman 2020.

Lenton, T. M.: *Early warning of climate tipping points, Nature climate change*, 1, 201–209, 2011.

Lenton, T., Livina, V., Dakos, V., and Scheffer, M.: *Climate bifurcation during the last deglaciation?*, Climate of the Past, 8, 1127–1139, 2012a. https://doi.org/10.5194/cp-8-1127-2012

Lenton, T. M., Livina, V. N., Dakos, V., van Nes, E. H., and Scheffer, M.: *Early warning of climate tipping points from critical slowing down: comparing methods to improve robustness,* Philosophical Transactions of the Royal Society A : 5 Mathematical, Physical and Engineering Sciences, A 370, 1185–1204,https://doi.org/10.1098/rsta.2011.0304, 2012b.

Lenton, T. M., Rockstrom, J., Gaffney, O., Rahmstorf, S., Richardson, K., Steffen, W., and Schellnhuber, H. J.: *Climate tipping points—too risky to bet against*, 2019.

Livina, V. N. and Lenton, T. M.: *A modified method for detecting incipient bifurcations in a dynamical system,* Geophysical Research Letters, 34, https://doi.org/https://doi.org/10.1029/2006GL028672, 2007.

Louchet, F., *Weather instabilities as a warning sign for a nearby climatic tipping point?* https://doi.org/10.48550/ARXIV.1609.05098, 2016.

Louchet, F., *How far can we trust climate change predictions?* arXiv:2204.11619v3 [physics.ao-ph], 2022 https://arxiv.org/abs/22 04.11619v3

Louchet, F., *Where the hell is climate change taking us? A clear and simple answer from Physics.* https://youtu.be/HmGfweH8YFw 2023 a

Louchet, F., *Climat : Evolution prévisible ou saut dans l'inconnu ?* Conference, Wissembourg, https://www.youtube.com/watch?v=D2ZNRH0o_D8 2023 b

Louchet, F., *Repeated Warning Signals for Sudden Climate Warming: Consequences on Possible Sustainability Policies.* Sustainability, *17*, 8548, 2025. https://doi.org/10.3390/su17198548

McInerney, F. A. and Wing, S. L.: *The Paleocene-Eocene Thermal Maximum: A Perturbation of Carbon Cycle, Climate, and Biosphere with Implications for the Future*, Annual Review of Earth and Planetary Sciences, 39, 489–516, https://doi.org/10.1146/annurev-earth-040610-133431, 2011.

Milesi, V., Guyot, F., Brunet, F., Richard, L., Recham, N., Benedetti, M., Dairou, J., & Prinzhofer, A., *Formation of CO_2, H_2 and condensed carbon from siderite dissolution in the 200–300 °C range and at 50 Mpa*, Geochimica & Cosmochimica Acta, 154, 201-211, 2015 .

https://doi.org/10.1016/j.gca.2015.01.015

Mlnlng.com, https://www.mining.com/what-may-be-the-worlds-largest-white-hydrogen-deposit-discovered-by-accident/

Neukom, R., Steiger, N., Gomez-Navarro, J. J., Wang, J., and Werner, J. P.: *No evidence for globally coherent warm and cold periods over the preindustrial Common Era*, Nature, 571, 550–554, 2019.

Pettit, E. C., Wild, C., Alley, K., Muto, A., Truffer, M., Bevan, S. L., Bassis, J. N., Crawford, A., Scambos, T. A., and Benn, D.: *Collapse of Thwaites Eastern Ice Shelf by intersecting fractures.*, in: AGU Fall Meeting 2021, AGU, 2021.

Saint-Martin, D., Geoffroy, O., Voldoire, A., Cattiaux, J., Brient, F., Chauvin, F., Chevallier, M., Colin, J., Decharme, B., Delire, C., Douville, H., Gueremy, J.-F., Joetzjer, E., Ribes, A., Roehrig, R., Terray, L., and Valcke, S.: *Tracking Changes in Climate Sensitivity in CNRM Climate Models*, Journal of Advances in Modeling Earth Systems, 13, e2020MS002 190, https://doi.org/https://doi.org/10.1029/2020MS002190,e2020MS002190 2020MS002190, 2021.

Steffen, W., Rockstrom, J., Richardson, K., Lenton, T. M., Folke, C., Liverman, D., Summerhayes, C. P., Barnosky, A. D., Cornell, S. E., Crucifix, M., et al.: *Trajectories of the Earth System in the Anthropocene*, Proceedings of the National Academy of Sciences, 115, 8252–8259, 2018. https://www.pnas.org/doi/pdf/10.1073/pnas.1810141115

St George, S.: *The aberrant global synchrony of present-day warming,* Nature. ;571(7766):483-484, 2019. doi: 10.1038/d41586-019-02179-2. PMID: 31341299.

Taleb, N. N, *The Impact of the Highly Improbable*, Penguin, 2007, 2010. * https://en.wikipedia.org/wiki/Black_swan_theory

Wild, C. T., Alley, K. E., Muto, A., Truffer, M., Scambos, T. A., and Pettit, E. C.: *Weakening of the pinning point buttressing Thwaites Glacier*, West Antarctica, The Cryosphere, 16, 397–417, https://doi.org/10.5194/tc-16-397-2022, 2022.

When the wheat is under hail

Crazy who does the delicate

Crazy who thinks about his quarrels

At the heart of the common fight

(extract from the poem La Rose et le Réséda by Louis Aragon)

Acknowledgements

This work would certainly have been different if it had not benefited from the assistance, encouragement, suggestions and proofreading of numerous colleagues and close friends, in particular Karim Azouaou, Dina Blanc, Jean Marie Casals, Michel Castel, Michèle Castelvallier, Xavier Chauvet, Karine Corbier, Michel Dupeux, John David Embury, Jérôme Faillettaz, Jérôme Garnier, Gilles Gauthier, Gerhard Krinner, Christopher D. Latham, Pierre Lelièvre, Jean Louchet, Marie France Louchet, Anne Louchet Chauvet, Cécile Louchet Dournel, Hubert Reeves and Erland M. Schulson.

Our warmest thanks to them.

Contact : f.louchet@gmail.com
scastel67@gmail.com
https://flouchet.wixsite.com/website

www.ingramcontent.com/pod-product-compliance
Lightning Source LLC
Chambersburg PA
CBHW051148220526
45473CB00003B/697
* 9 7 9 8 3 4 0 3 0 9 8 9 1 *